The Pendulum Workbook

18.95

The Pendulum Workbook

BOTE MIKKERS

Translated from the Dutch by
HANS HOEKMAN

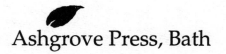
Ashgrove Press, Bath

Published in Great Britain by
ASHGROVE PRESS LTD
7 Locksbrook Road Estate
Bath, BA1 3DZ

Originally published in Dutch by
Ankh-Hermes, Deventer

ISBN 1-85398-036-6

First published in English 1994

Typeset by Solos Press
Motcombe, Dorset
Printed in England by
Redwood Books, Trowbridge

Contents

Introduction 7

1. Astrology 13
2. Numbers, calculations and letters 22
3. Archaeology 28
4. The causes of disease 34
5. Localization of diseases 38
6. Treatments 57
7. Vitamins and minerals 61
8. Homoeopathy 67
9. Gemstone therapy 84
10. Radiation 90
11. Allergy 96
12. The aura 105
13. The chakras 108
14. I Ching 112
15. Weather 116
16. Holidays 119
17. Sports and leisure 128
18. Character traits, areas for special attention 132
19. Headlines 143
20. Minor decisions, Important issues 150
21. Professions and business 155
22. Lost objects and missing persons 169

Introduction

Just like rod-divining, pendulum dowsing is an age-old custom. Through the years, a great many books have been written on this subject. Some of these books were excellent, others were of rather poor quality. To obtain the right information may, therefore, prove to be somewhat problematic for someone who is seriously investigating this phenomenon. Some of the questions which may arise are:
- Can anyone use the pendulum?
- Can you learn how to use it?
- How do you find out whether you are 'sensitive' or not?
- Can the pendulum be used under all circumstances?
- Are there any risks involved? If so, what are they?

In this introduction I will explain how to use the pendulum and how it works. Also, the above-mentioned questions will be discussed in more detail. When you use the pendulum it is of the utmost importance - as it is in dowsing - to determine very clearly beforehand how 'yes' or 'no' is indicated; what is positive, and what is negative. You can determine this yourself. For instance, you may consider a swinging movement, comparable to shaking your head, as negative, and an opposite movement like a nod, as positive. It is also possible to designate a circular movement to the left as 'no' (negative), and a circular movement to the right as 'yes' (positive).

If the pendulum hangs still, this may mean that the question was not formulated correctly; in this case the answer is neither 'yes' nor 'no', neither positive nor negative. When asking a question, it is assumed that you ask something to which you do not yet know the answer, which nevertheless is important to you. Please do not ask any unnecessary questions. One should use the pendulum in a responsible way, also from an ethical point of view. This means that one should not use the pendulum to obtain information which is used - or could be used - wrongly. Personally, I never use the pendulum to get information on people who would certainly not have given me permission to do so. I would refuse to use the pendulum, for instance, if a suspicious spouse wanted to know whether her husband had any extra-marital relations. Neither very meaningful are questions like: What does my future partner look like? Where and when shall I meet him/her? On the other hand, questions from worried parents whose child is ill or missing, I take very seriously. Only use the pendulum to help yourself and others in the proper sense of the word, without harming the person that you are inquiring about. As stated earlier, the pendulum is used only to ask a question about an existing situation to which the answer is yet unknown. This question formulated in our conscious mind passes through the subconscious into the unconscious.

Our immediate perception of everything is registered in our conscious mind. From this domain of direct knowledge, the perception sinks into the subconscious, from which we can retrieve it as direct memory. Information from the subconscious also reaches the domain of the unconscious. Here, knowledge is also present, only we are not aware of it. It is possible to become aware of this knowledge by means of clairvoyance, clairaudience, precognition, intuition, and smell. We all can draw from this unconscious source of knowledge. This is why it is called the 'collective unconscious'. The unconscious gets impressions from all kinds of places and in many ways, also without our direct consciousness. The experiences of many centuries and many lives are recorded here. So, using the pendulum, one asks questions which one simply cannot answer on the basis of the information available in our direct consciousness. Yet, these questions can be answered with the help of the information that is stored in the unconscious. When in earnest and without prejudice, one may receive answers by means of one's unconscious, brought about through muscle reflexes stemming from the autonomic nervous system. To the question whether anyone can do this, the answer is simple: there are just as many people who can learn to use the pendulum, as there are who can learn to play the piano. Some are gifted and have easy access to their unconscious, others are not that fortunate. Whether information is reliable and valuable largely depends on how gifted a person is. Another important factor is one's attitude. Let us compare pendulum dowsing to playing a musical instrument. A great talent may wither without practice, whereas someone less talented may achieve success through concentrated study and self-control. A very gifted person may no doubt achieve even more by applying himself. Some people make good musicians, but will never become concert pianists. Similarly, not everyone will excel in divining, pendulum-dowsing or clairvoyance.

This workbook is meant to encourage people to develop a possible talent for pendulum dowsing. A talented pendulum user, clairvoyant, or psychic will be able to determine whether - and to what degree - you are 'sensitive'. You may find out for yourself by doing blind tests with a dowsing rod or pendulum.

These tests should be prepared by someone else. The following test may serve as an example. Take ten similar boxes and let someone else put objects in two or three boxes. It is then up to you, using your pendulum or dowsing rod, to determine in which of the boxes an object is hidden.

These tests can be extended and are applicable in various fields. For instance, if someone has a certain disability or illness, you may be able to locate the disturbance, diagnose what is wrong, and check whether your diagnosis was correct.

When do you use the pendulum?

Basically, you can use the pendulum for any purpose. But it is certainly not recommended to use the pendulum for egoistic and materialistic purposes. This will not work and is a waste of energy. If your attitude is wrong, you get an incorrect answer. If you should want to use the pendulum to win the lottery, this would only prove to be a complete waste of time. The various ways of using the pendulum are all described further down.

Can the pendulum be used under all circumstances?

Most certainly not. It only makes sense to use the pendulum if you are in total harmony with yourself and the world around you. It would not work, for example, right after you had had an argument with someone. If you want to use a specific gift, such as pendulum dowsing, wisely, then you should observe the following rules:
- You should not cherish negative feelings such as hate,jealousy, or suspicion.
- You should not use the pendulum, if you cannot receive the   answer with an open mind.
- You must be sound in mind and body.
- You should not use the pendulum for matters that do not concern you.
- You should not use the pendulum for yourself or others, if   the information will be used negatively or will harm someone.
- You should not pass on any information to third parties, if those concerned are not consulted first.
- You should not use the pendulum in an environment in which the harmony is disturbed, e.g. because of loud music or quarrels in the background.
- You should not use the pendulum to prove your skill or to satisfy someone else's curiosity or thirst for sensation.
- You should not use the pendulum casually. Take your time and do it right.

Are there any risks involved?

Every activity has its risks, and pendulum dowsing is no exception. Nobody is perfect. The perfect dowser does not exist. If you accept the answers you receive without questioning and act accordingly, you may have to suffer the consequences. The conscious and unconscious need to be in tune, need to form a unity. Answers from the unconscious which you receive by means of clairvoyance, clairaudience, precognition, intuition, and pendulum dowsing should be harmonized and form a unity with the conscious domain of knowledge. It is only then that they become of use. Regularly, you meet people who claim that they use the pendulum to receive answers from their spiritual guides. Again others receive messages from the dead, spirits or entities from other worlds or spheres of existence. I would like to stress here that all these answers come from these people's own unconscious. To what extent our unconscious is attuned to spiritual entities and what character these have, depends largely on our own spiritual and mental level. If the answers you receive from your unconscious by means of the pendulum are influenced by certain entities, the reliability of the information will be proportionate to your own inner motivation. The risk involved in pendulum dowsing is that wrong motives lead to unreliable answers. In such case you cheat yourself and others, which only makes matters worse.

What pendulum to use

The pendulum is an object made of metal, wood, glass, diamond, or any other material fixed to a thread or a small chain (fig.1). If you want to work accurately, it is advisable to use a round, somewhat pointed shape. It does not matter what material you use, or whether you use a thread or a small chain. As mentioned earlier, the most important thing is the muscle reaction which guides the hand, via the unconscious through the subconscious and autonomic nervous system.

The ideal length of the thread is up to you. In general, this is somewhere between ten and twenty centimetres. It is advisable to fit the end of the chain or string with a knot, so that it will not slip from your fingers. There are all sorts of pendulums, as far as material and design are concerned. Choose the material and shape you like. All pens are made to write with, but you can choose between ball pens, roller pens and fountain pens in different colours, sizes and shapes. Normally, you choose the colour and model you like best. That is the one you like to work with. It gives you an extra impetus, which is very important. Figure 1 represents various kinds of pendulums.

An Example

With the use of figure 2 you can obtain information about wind-direction, degrees, blood pressure, and intelligence quotient. For example, you may want to determine the wind-direction before you go on a sailing or cycling trip.

Hold the pendulum above the centre of the diagram and ask: From which direction will the wind blow? The pendulum swings at an angle of 45 degrees from top right to bottom left. All you now have to do is to determine whether the answer is 'top right' or 'bottom left'. Next, hold the pendulum over NE and ask: Is this the right direction? If the pendulum indicates 'yes' by swinging clockwise, you know the wind-direction will be northeast.

In the same manner you are able to determine your blood pressure. For example, first ask: What is the systolic pressure?

Now suppose that from the centre of the diagram the pendulum once again starts to swing from bottom left to top right over the numbers 140 and 320. Hold the pendulum over number 320 and ask: Is this the value for the systolic pressure?

If the pendulum swings anti-clockwise and indicates 'no', you know that number 140 is the value for the systolic pressure. To determine the diastolic pressure, you follow the same procedure.

You do likewise in order to determine degrees or the intelligence quotient.

With a circle diagram, always check which is which, as there are always two possibilities.

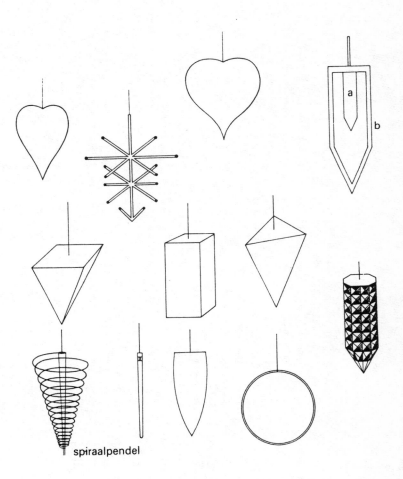

spiraalpendel

1. Astrology

Let us assume that you would like to cast someone's horoscope and need to know the date and time of birth. In order to determine these, you can make use of the diagram (fig.3.1).

You may use the following procedure. Hold the pendulum approximately 5-10 mm above the black spot in the centre of figure 3.1. Then, ask in which month the person in question was born. Let us assume that the pendulum starts to swing from top right to bottom left along the line marked 'September'.

You then know that the month was September. Next, you would like to know which day in September this person was born. Hold the pendulum directly above the black spot of figure 3.2, and ask your pendulum to indicate the exact date. The pendulum may start swinging, for example, almost horizontally with a slight deviation to top right and bottom left. The pendulum is swinging over the section marked 'see above'. This means that you have to repeat the question using figure 3.1 instead. Doing so, the pendulum again moves almost horizontally, but now it is deviating slightly to top left and bottom right. The pendulum indicates '23'. Hence, the person in question was born on 23rd September.

To ascertain which day of the week it was, hold the pendulum over figure 3.3 and formulate your question. It turns out the pendulum indicates 'Friday'.

Subsequently, you would like to know the time of birth. Hold the pendulum over figure 3.2, again directly above the black spot, and ask: At what time was this person born? The pendulum now may,for example, start to swing almost vertically, deviating slightly to top left and bottom right. The time indicated is eleven o'clock.

Hold the pendulum at the bottom of figure 3.4 and ask whether it was exactly eleven o'clock or some minutes earlier or later. The pendulum swings to top right, so you know that some minutes need to be subtracted. To find out how many minutes, hold the pendulum over figure 3.2 and ask how many minutes before eleven o'clock this person was born. Assuming that the pendulum again indicates 'see above', turn to figure 3.1 and repeat your question. Now, 25 minutes are indicated.

You have found that the person in question was born on 23rd September at 10.35 am, i.e. eleven o'clock minus 25 minutes. If you would also like to know whether the year is correct - or if you do not know the year - continue with the diagrams on the next page.

Let us use the example above. You know that this person was born on Friday 23rd September, but you do not know in which year. Hold the pendulum over the diagram marked 'Orientation' of figure 4, directly above the point where all the lines meet and ask: In which diagram can I find the correct year? The swing of the pendulum indicates figure 4.2 . Turn to figure 4.2 and ask: In which year was this person born? The pendulum indicates the year 1938.

Now you know that the person in question was born in 1938, on Friday 23th September at 10.35 am, which turns out to be most fortunate, given the favourable opportunities and prospects.

If you would like to know the place of birth, it is advisable to use the diagrams in chapter 16 on 'Holidays'. With the help of the diagrams 'Parts of Britain' (fig.84), you will be able to determine the specific area. For the exact place of birth you need a good map of this area. Using this map, proceed as follows. Hold the pendulum at the bottom of the map, somewhere near the middle, and ask: Where was this person born? The pendulum will start to swing in a straight line. Then, hold the pendulum at the side of the map. Repeat the question. The pendulum will again swing in a straight line and this line will cross the previous one. At the point where the two lines cross, the person in question was born.

At the end of this chapter you will find some diagrams to determine star signs and the position of the planets.

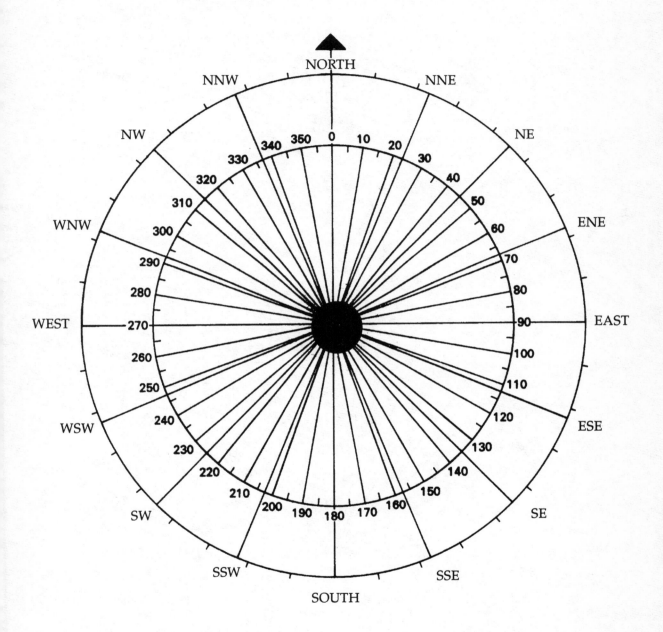

Figure 2. Wind-direction, degrees, blood pressure, intelligence quotient

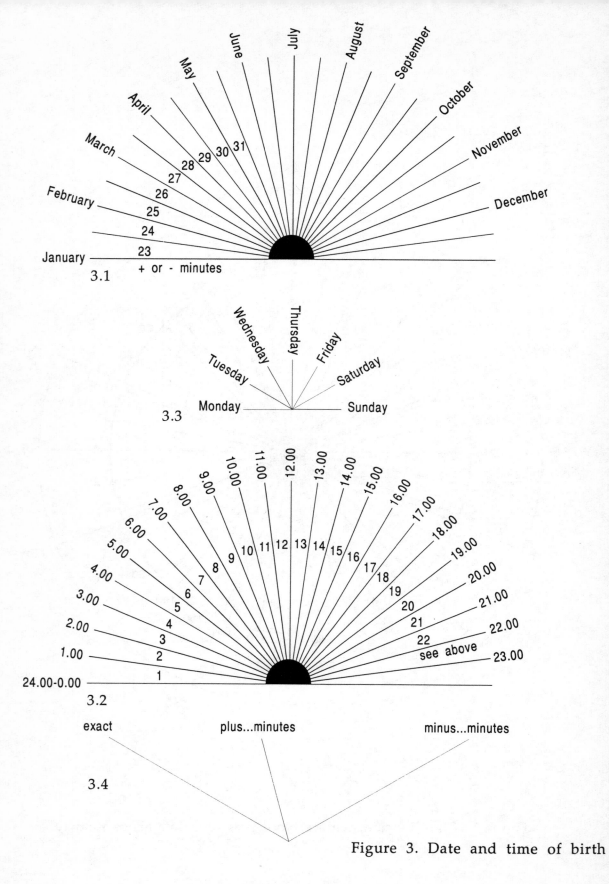

Figure 3. Date and time of birth

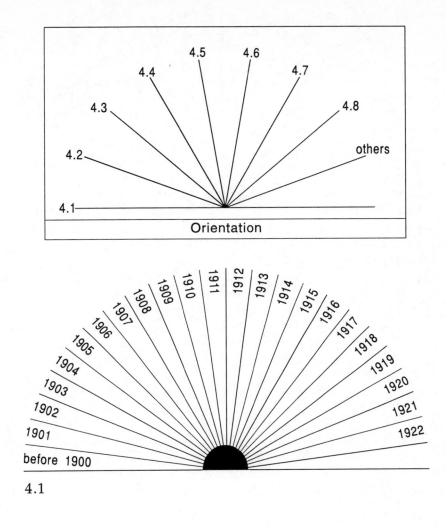

4.1

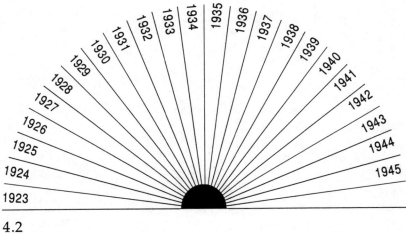

4.2

Figure 4. Years (1900-2080)

17

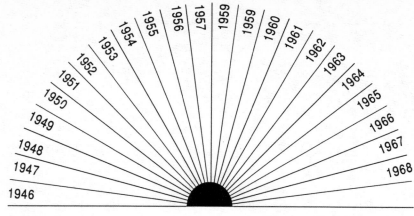

4.3

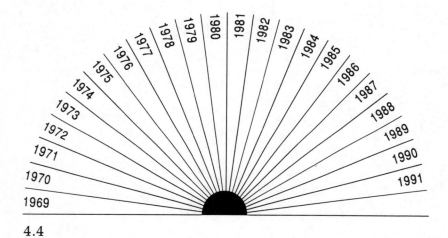

4.4

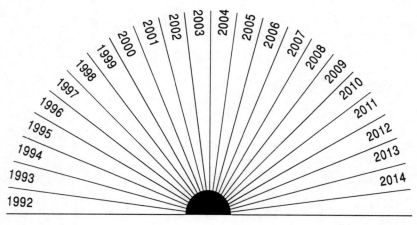

4.5

18

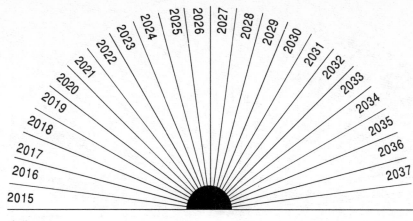

4.6

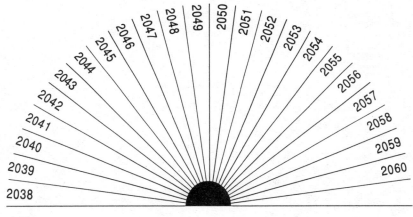

4.7

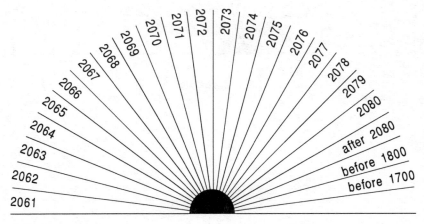

4.8

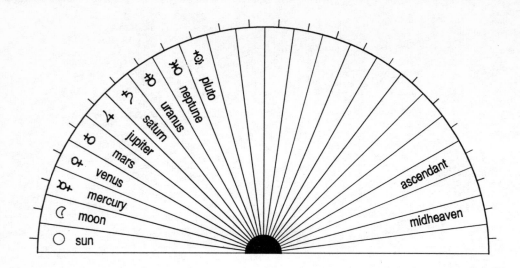

Figure 5. Planet

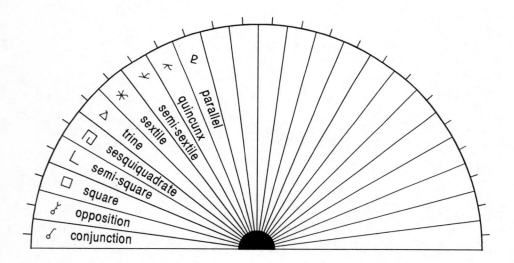

Figure 6. Aspect

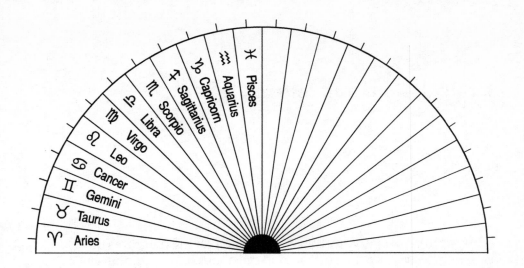

Figure 7. The zodiac

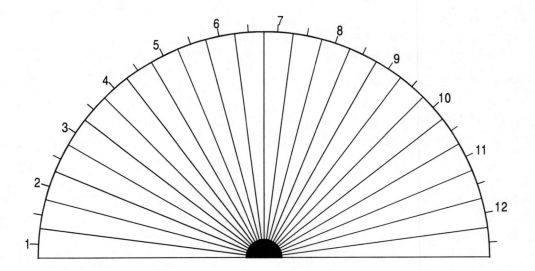

Figure 8. Number

2. Numbers, calculations and letters

Numbers are a regular feature in everyday life. With the help of the diagrams in this chapter, it is possible to use the pendulum to find out about house numbers, the pages in a book which contain a certain passage, lost numeric codes, certain calculations, number plates, postal codes, and many other things.

It is important that you start at the diagram marked 'Orientation'. Here, you ask in which diagram you will find your answer.

On the next pages you will find some diagrams with high numbers. These can be used for many purposes, not just for dates. These numbers are multiples of 100. If you would like to know whether you have the right number or that something has to be added or subtracted, use the diagram with 'higher' and 'lower' (fig.9.8). Subsequently, use the diagrams with the lower numbers.

For the years after 1900 and exact dates (day,week,month), use the diagrams from chapter 1 on 'Astrology'.

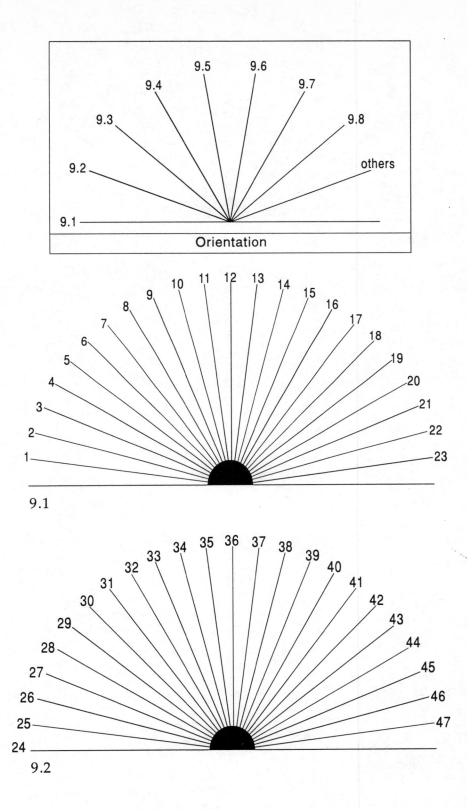

Figure 9. Numbers and calculations

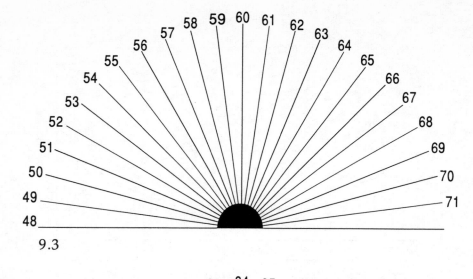

9.3

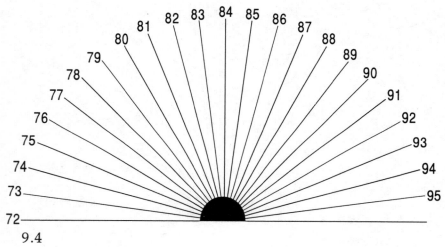

9.4

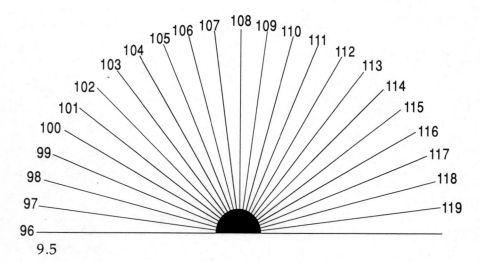

9.5

24

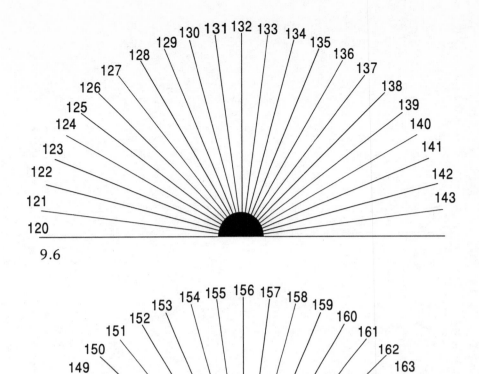

9.6

9.7

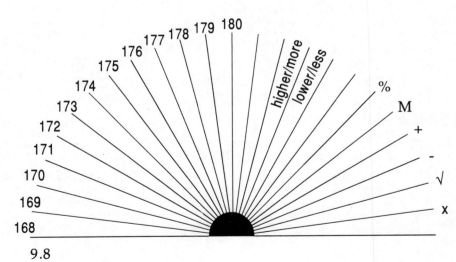

9.8

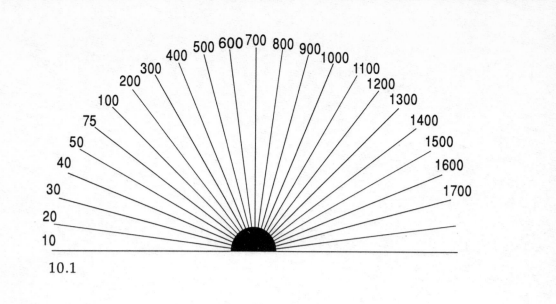

10.1

exact

more   less

10.3

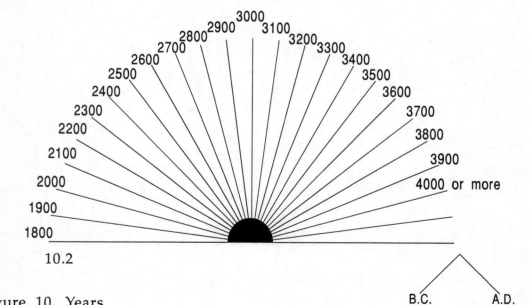

10.2

10.4

B.C.   A.D.

Figure 10. Years

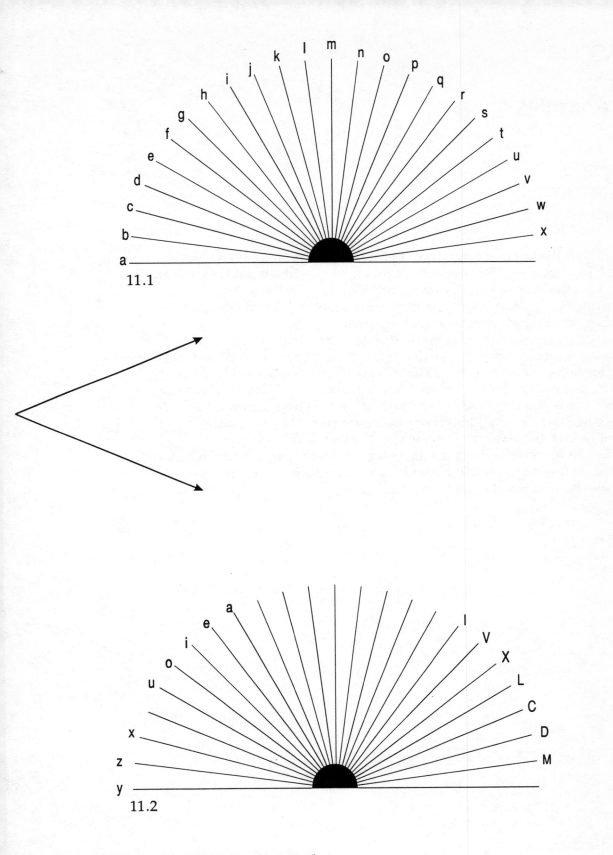

11.1

11.2

Figure 11. Letters and Roman numerals

3. Archaeology

Much of our distant and more recent past still lies hidden under the earth's surface. The following diagrams are meant for people who would like to know more about the lives of our ancestors, and what they left behind.

To date objects, please refer to chapter 2 on 'Numbers, calculations and letters'. There you will find a diagram containing dates.

A detailed map is also very helpful to locate archaeologically interesting sites. If this map is rather large, you can divide it into four or more sections. Holding the pendulum over each section, ask: Is there any archaeologically interesting object in this section (or the object you are looking for)? The pendulum will indicate 'yes' or 'no'. Once you have found the section, you can proceed further. To this end, hold the pendulum above the bottom line of the section and ask the pendulum to indicate the direction where you may find the object.

Subsequently, hold it above the left or right side of the section and repeat the question. In this way you will have two lines that intersect at a given point on the map. This is the spot to start digging.

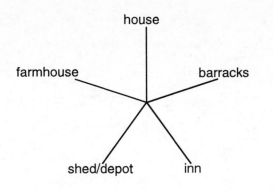
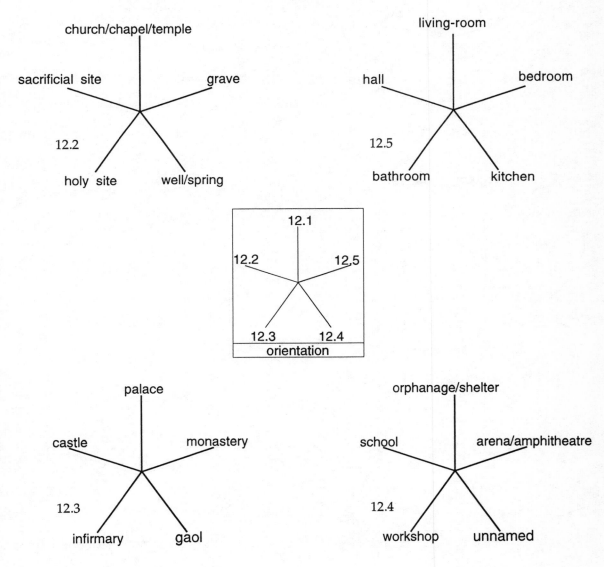

Figure 12. Archaeological remains of buildings

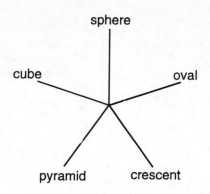

sphere

cube   oval

pyramid   crescent

13.1

triangle

diamond   square

13.2

parallelogram   trapezium

13.5

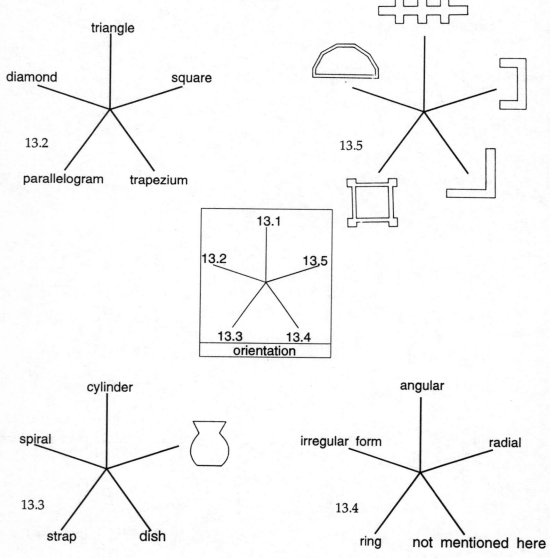

13.1

13.2   13.5

13.3   13.4

orientation

cylinder

spiral

13.3

strap   dish

angular

irregular form   radial

13.4

ring   not mentioned here

Figure 13. Shape of objects

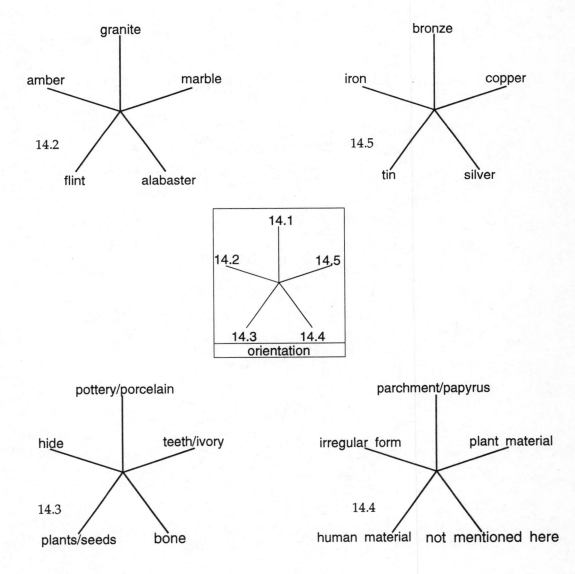

Figure 14. Materials

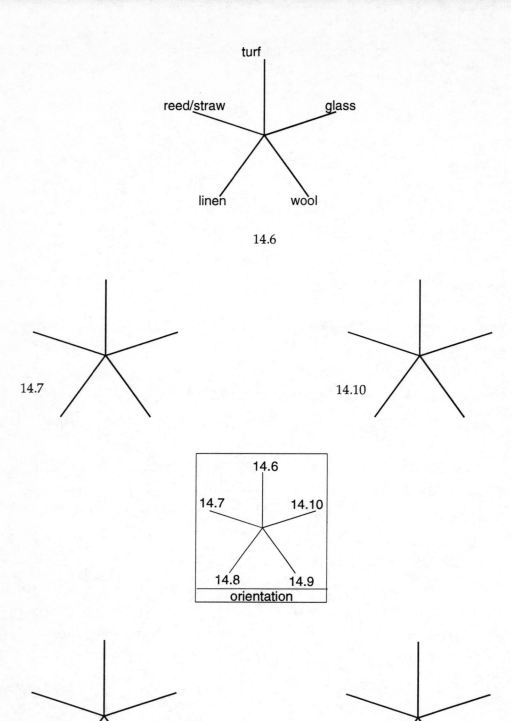

turf

reed/straw   glass

linen   wool

14.6

14.7

14.10

14.6

14.7   14.10

14.8   14.9

orientation

14.8

14.9

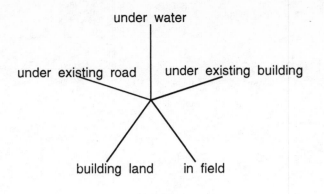

under water

under existing road | under existing building

building land | in field

15.1

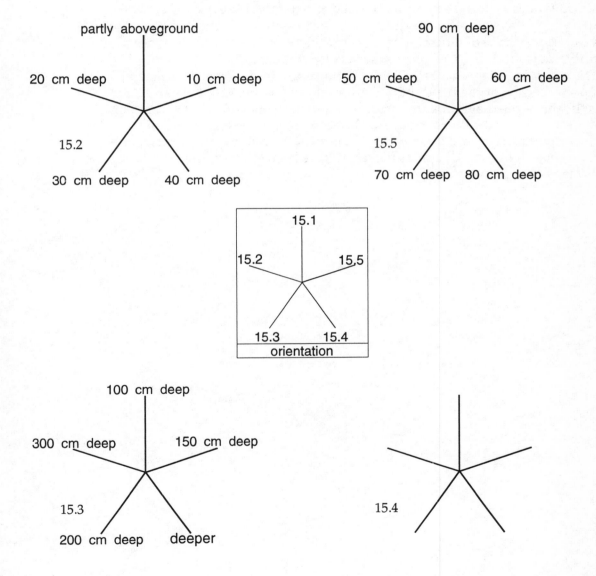

partly aboveground

20 cm deep | 10 cm deep

15.2

30 cm deep | 40 cm deep

90 cm deep

50 cm deep | 60 cm deep

15.5

70 cm deep | 80 cm deep

15.1

15.2 | 15.5

15.3 | 15.4

orientation

100 cm deep

300 cm deep | 150 cm deep

15.3

200 cm deep | deeper

15.4

Figure 15. Sites

4. The causes of disease

Health is a state in which body, soul and mind are balanced. The body will always try to preserve this state. A severe disturbance or stress may jeopardize the balance and can be the cause of illness. A great many disturbances can occur. Once the disturbance is localized, treatment may become more effective.

The following diagrams provide the opportunity to discover the cause of a disease by means of the pendulum. Start each time by holding the pendulum directly above the diagram in the middle of the page and ask: Which diagram shows the cause of the illness from which this person suffers? Mention his/her name, or possibly, attune yourself to this person by using a photograph or a personal object. In some diagrams the terms have been omitted, so that you can fill these in yourself.

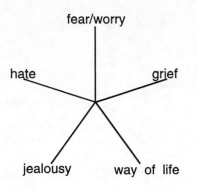

fear/worry

hate   grief

jealousy   way of life

16.1

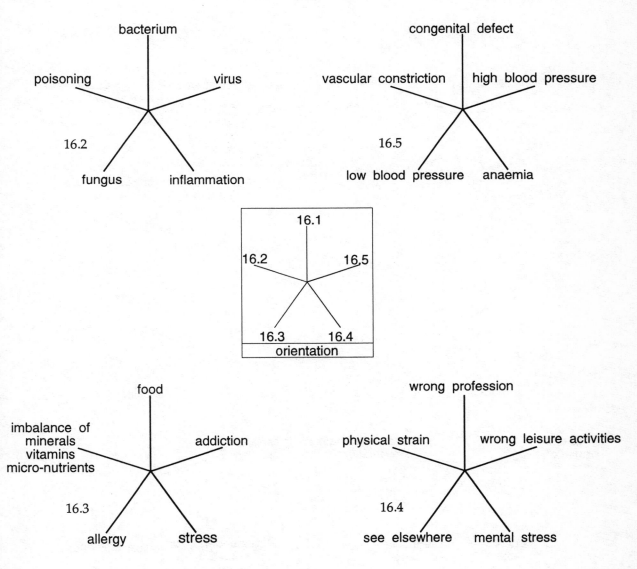

bacterium

poisoning   virus

16.2

fungus   inflammation

congenital defect

vascular constriction | high blood pressure

16.5

low blood pressure   anaemia

16.1

16.2   16.5

16.3   16.4
orientation

food

imbalance of
minerals
vitamins   addiction
micro-nutrients

16.3

allergy   stress

wrong profession

physical strain   wrong leisure activities

16.4

see elsewhere   mental stress

Figure 16. Causes of ill health

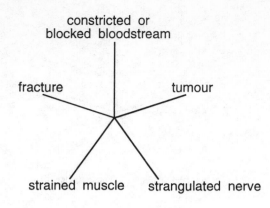

constricted or
blocked bloodstream

fracture   tumour

strained muscle   strangulated nerve

16.6

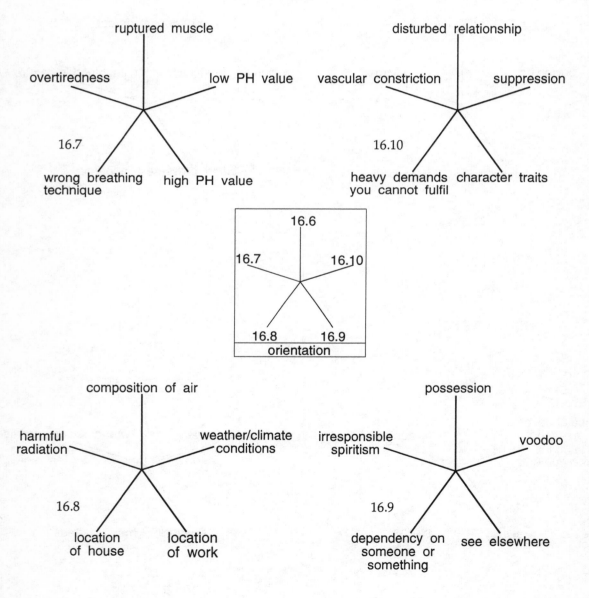

ruptured muscle

overtiredness   low PH value

16.7

wrong breathing   high PH value
technique

disturbed relationship

vascular constriction   suppression

16.10

heavy demands   character traits
you cannot fulfil

16.6

16.7   16.10

16.8   16.9
orientation

composition of air

harmful   weather/climate
radiation   conditions

16.8

location   location
of house   of work

possession

irresponsible   voodoo
spiritism

16.9

dependency on   see elsewhere
someone or
something

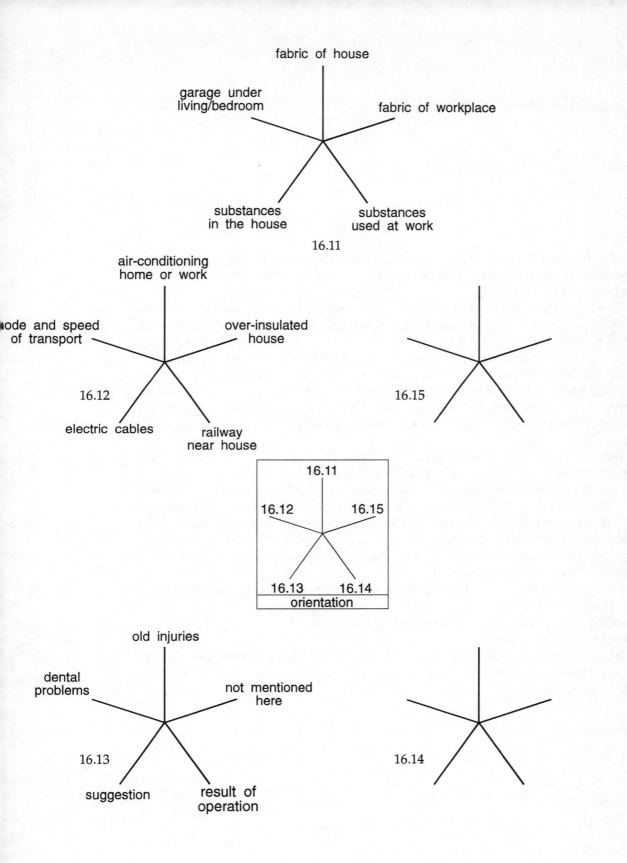

fabric of house

garage under
living/bedroom

fabric of workplace

substances
in the house

substances
used at work

16.11

air-conditioning
home or work

mode and speed
of transport

over-insulated
house

16.12

electric cables

railway
near house

16.15

16.11

16.12   16.15

16.13   16.14
orientation

old injuries

dental
problems

not mentioned
here

16.13

16.14

suggestion

result of
operation

5. Localization of diseases

On the following pages you will find some diagrams that will help you to determine from which disease a person is suffering, or which organs or parts of the body are troubled.

There are three pages that show diagrams with groups of diseases. Please proceed as follows. Hold the pendulum directly above the centre of the diagram in the middle of the page and ask: In which section can I find the class of disease from which this person is suffering? State name and, possibly, time and place of birth; a photograph or personal object, such as a piece of jewellery, may also help. The pendulum will then indicate the class of disease. Next, hold the pendulum above the relevant section, and ask: what type of disease/disability does this person suffer from? The pendulum will now indicate more specifically what type of disease is present.

Figure 17 represents the human body. This may help you to find the organs or parts of the body that are of importance. Start at the orientation point and ask the pendulum to locate the area which causes problems. The pendulum will indicate one of the diagrams. Hold the pendulum directly above the black spot of that diagram and formulate your question. The pendulum will point out the specific organ or part of the body which causes the problems. On the next two pages you will find two diagrams of figure 18. These are simplified iris charts in which all organs are represented as well. Start off by asking whether the organ can be found on the left or right hand page. Hold the pendulum above the black centre of the page and ask it to indicate the affected organ of the person in question. The pendulum will swing to and fro, it is therefore important to ask which side is the right one.

In the last pages of this chapter detailed diagrams show a further specification of physical organs. These will enable you to determine what exactly is wrong with the organ.

1. cerebrum, nerve centre
2. cerebral membrane
3. eyes
4. ears
5. nasal bone
6. jaw
7. tonsils
8. larynx
9. thyroid gland
10. windpipe
11. shoulder joint
12. bronchi
13. right lung
14. liver front right
15. gall baldder, beh

16. pancreas, behind stomach
17. large intestine, transverse colon
18. abdominal aorta
19. inferior vena cava
20. large intestines, ascending colon
21. large intestines, descending colon
22. small intestine, ileum
23. caecum, front right

24. appendix
25. glands
26. fingernails
27. blood circulation, arteries, veins
28. veins
29. toenails
30. hair
31. pineal gland
32. small brain
33. spinal cord
34. aorta
35. apex of the lung
36. thymus gland
37. left lung
38. oesophagus
39. heart, betwen lungs
40. stomach, behind liver
41. elbow
42. spleen
43. kidneys
44. pelvis
45. urethra
46. hip joint
47. bladder
48. wrist joint
49. sex organs
50. knee joint
51. ankle joint

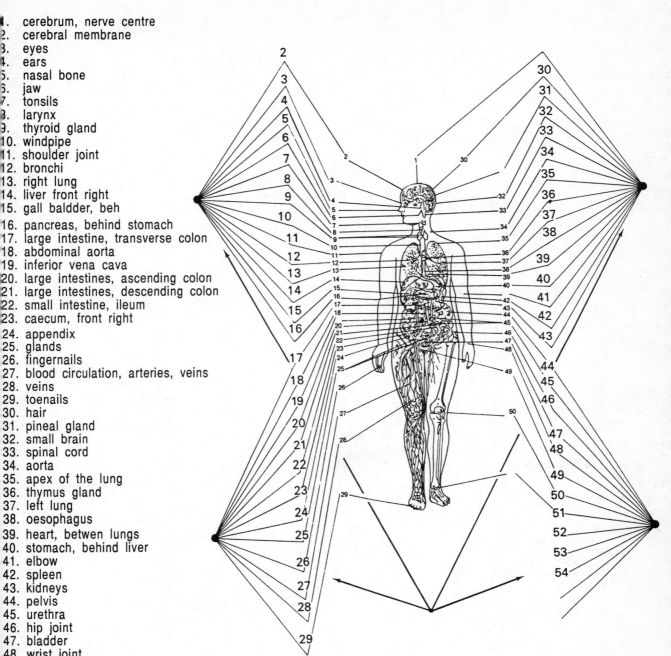

figure 17. The human body

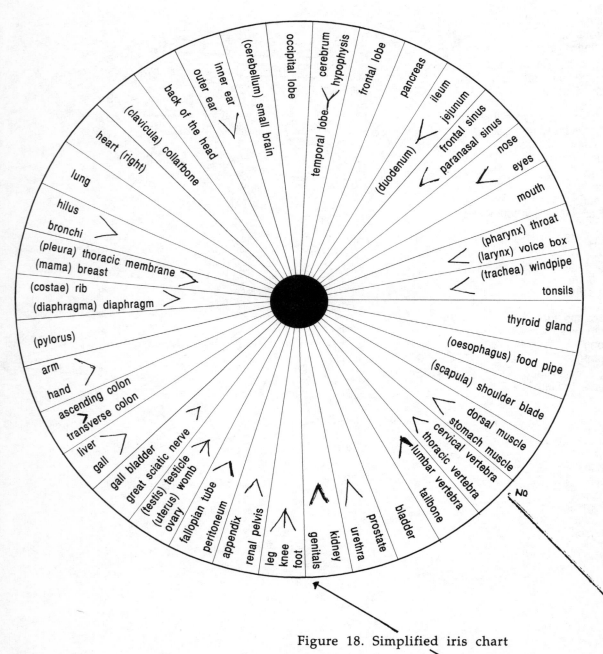

Figure 18. Simplified iris chart

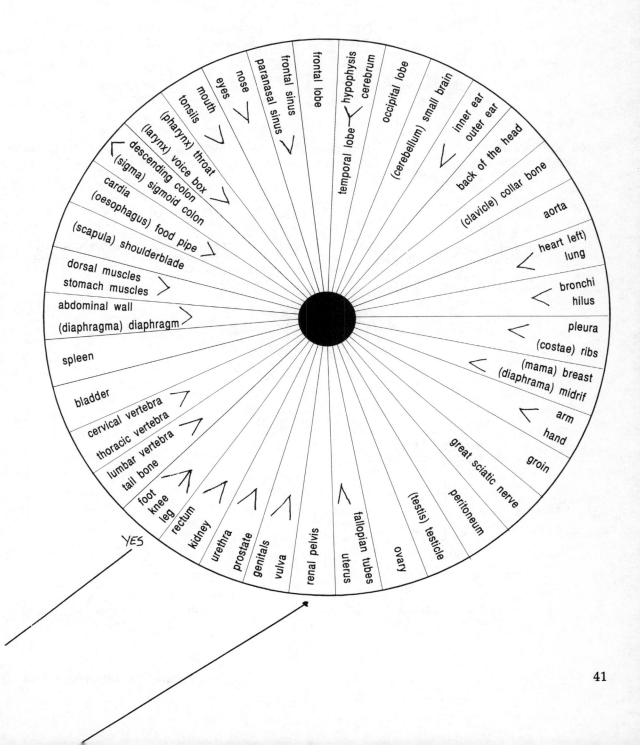

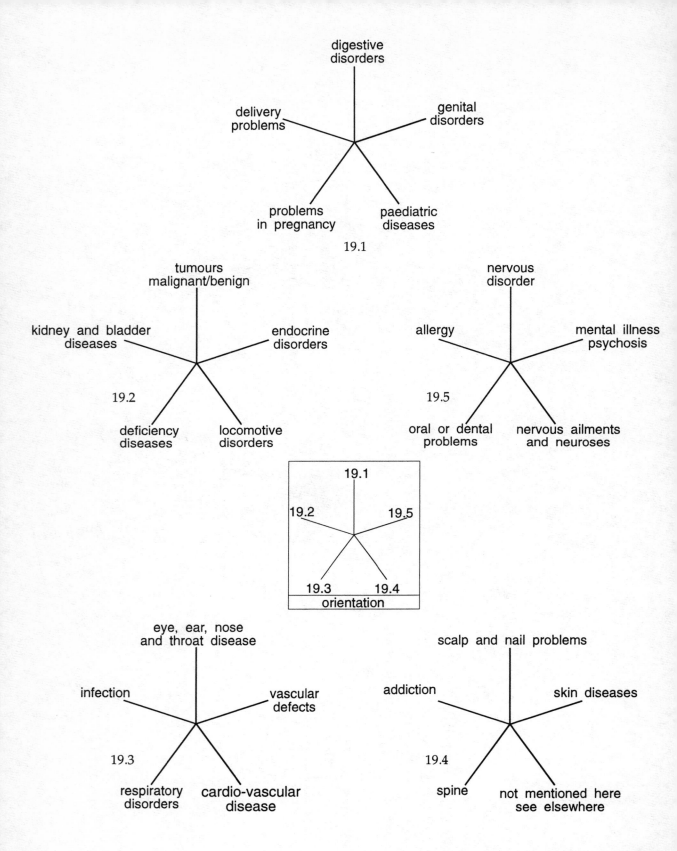

digestive
disorders

delivery
problems

genital
disorders

problems
in pregnancy

paediatric
diseases

19.1

tumours
malignant/benign

kidney and bladder
diseases

endocrine
disorders

19.2

deficiency
diseases

locomotive
disorders

nervous
disorder

allergy

mental illness
psychosis

19.5

oral or dental
problems

nervous ailments
and neuroses

19.1

19.2

19.5

19.3

19.4

orientation

eye, ear, nose
and throat disease

infection

vascular
defects

19.3

respiratory
disorders

cardio-vascular
disease

scalp and nail problems

addiction

skin diseases

19.4

spine

not mentioned here
see elsewhere

42

Figure 19. Disorders and defects

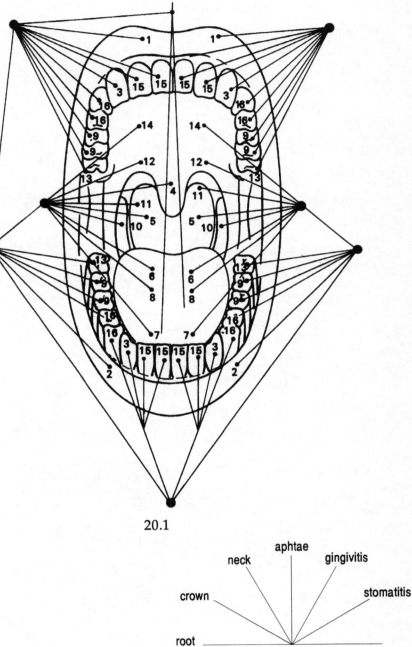

1. upper lip
2. lower lip
3. canine teeth
4. uvula
5. pharyngeal isthmus
6. vallate papillae
7. filiform papillae
8. fungiform papillae
9. molars
10. palatine tonsil
11. palatine arch
12. rear palate
13. wisdom teeth
14. upper jaw
15. incisors
16. premolars

20.1

neck   aphtae   gingivitis
crown   stomatitis
root

20.2

figure 20. parts of the mouth

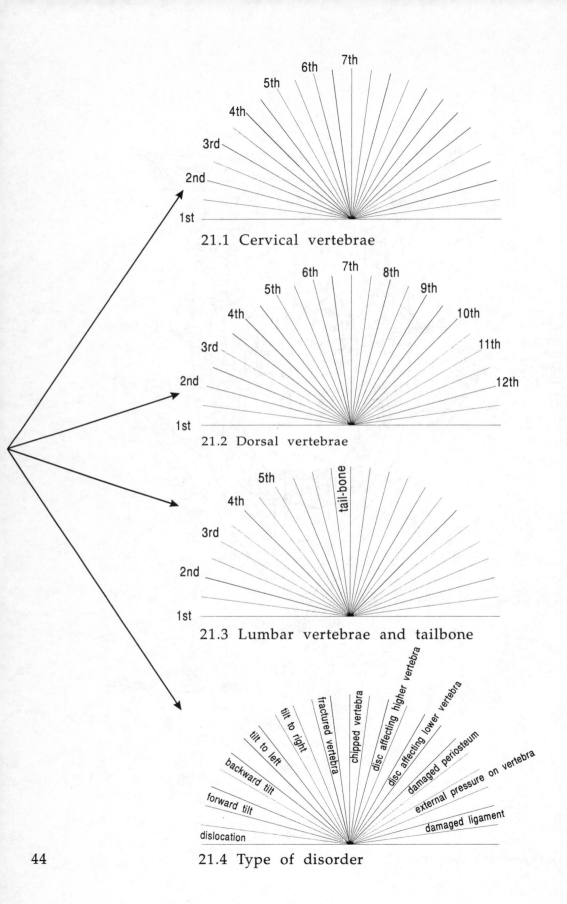

7th

6th

5th

4th

3rd

2nd

1st

21.1 Cervical vertebrae

6th 7th 8th

5th 9th

4th 10th

3rd 11th

2nd 12th

1st

21.2 Dorsal vertebrae

5th

4th tail-bone

3rd

2nd

1st

21.3 Lumbar vertebrae and tailbone

tilt to right

fractured vertebra

chipped vertebra

disc affecting higher vertebra

tilt to left

disc affecting lower vertebra

backward tilt

damaged periosteum

external pressure on vertebra

forward tilt

damaged ligament

dislocation

21.4 Type of disorder

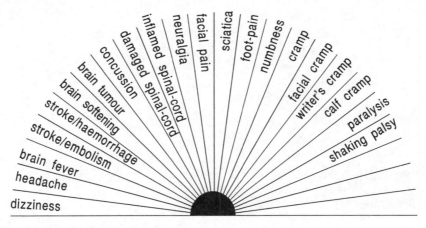

Figure 22. Disorders of the nervous system

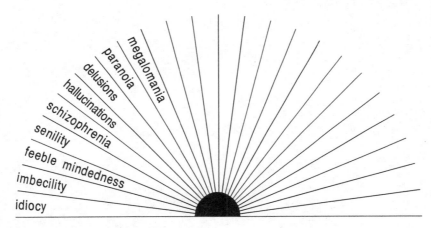

Figure 23. Mental illness and psychoses

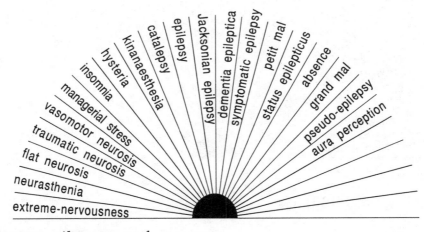

Figure 24. Nervous ailments and neuroses

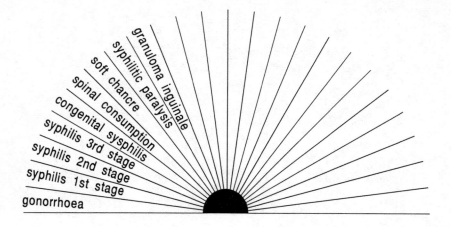

Figure 25. Venereal diseases

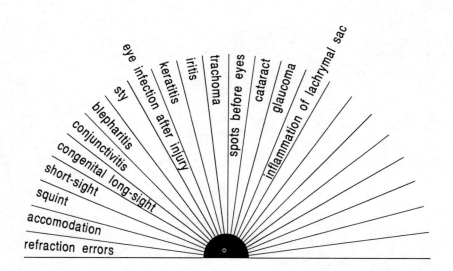

Figure 26. Eye troubles

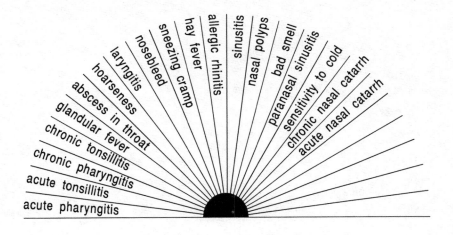

Figure 27. Pharyngeal and nasal disorders

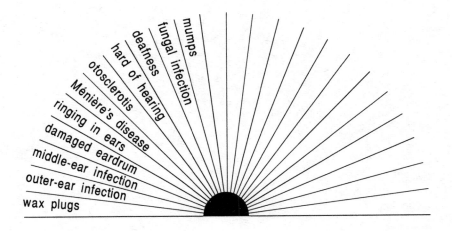

Figure 28. Ear troubles

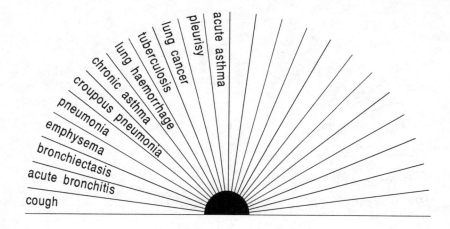

Figure 29. Respiratory disorders

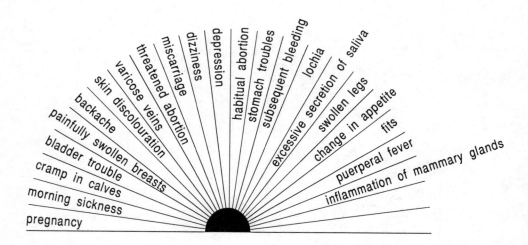

Figure 30. Pregnancy problems

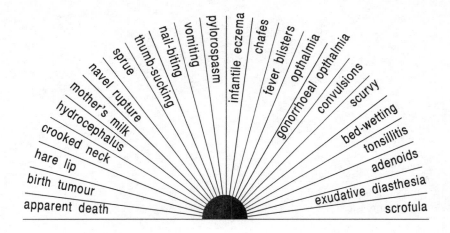

Figure 31. Physical defects and pediatric diseases

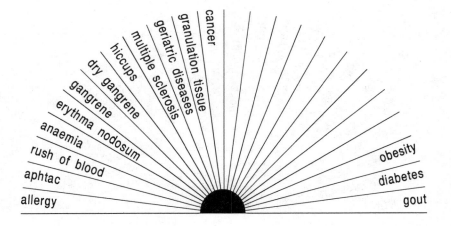

Figure 32. Common disorders and specific defects;
metabolic disorders

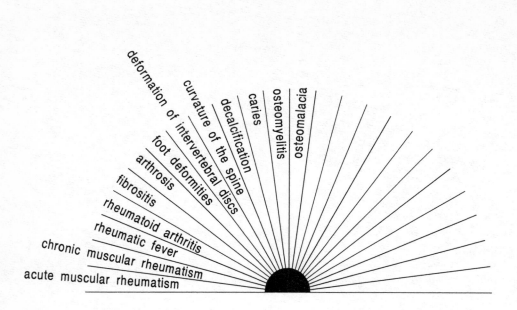

Figure 33. Diseases of the locomotive system; bone diseases

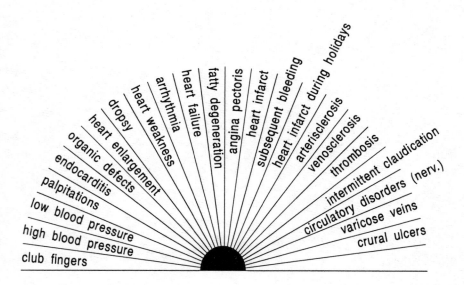

Figure 34. Cardiovascular diseases

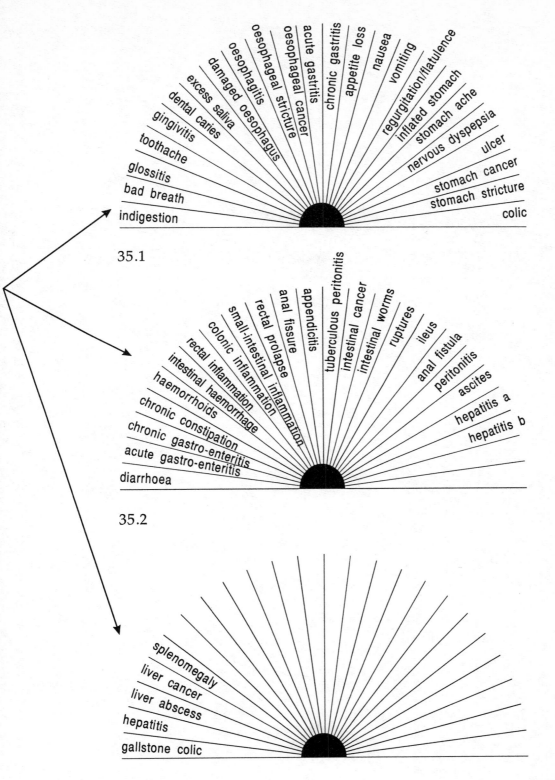

35.1

oesophageal cancer
oesophageal stricture
acute gastritis
chronic gastritis
appetite loss
nausea
vomiting
regurgitation/flatulence
inflated stomach
stomach ache
nervous dyspepsia
ulcer
stomach cancer
stomach stricture
colic
oesophagitis
damaged oesophagus
excess saliva
dental caries
gingivitis
toothache
glossitis
bad breath
indigestion

35.2

tuberculous peritonitis
anal fissure
appendicitis
intestinal cancer
rectal prolapse
intestinal worms
small-intestinal inflammation
ruptures
colonic inflammation
ileus
rectal inflammation
anal fistula
intestinal haemorrhage
peritonitis
haemorrhoids
ascites
chronic constipation
hepatitis a
chronic gastro-enteritis
hepatitis b
acute gastro-enteritis
diarrhoea

35.3

splenomegaly
liver cancer
liver abscess
hepatitis
gallstone colic

Figure 35. Digestive disorders

51

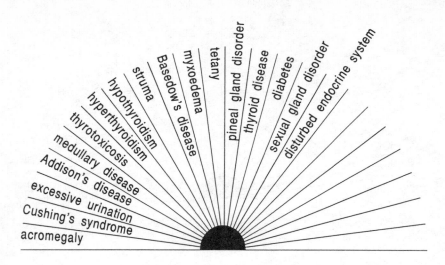

Figure 36. Diseases of the endocrine system

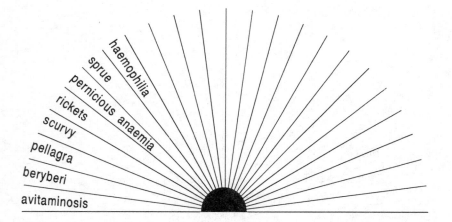

Figure 37. Deficiency disease

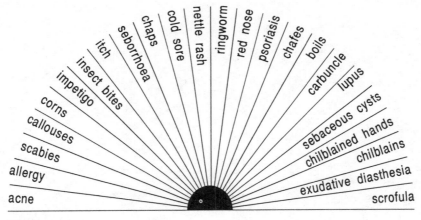

38.1

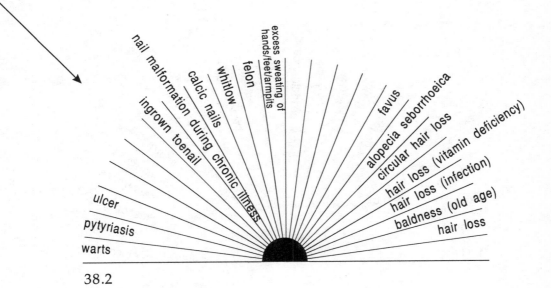

38.2

Figure 38. Skin disorders

53

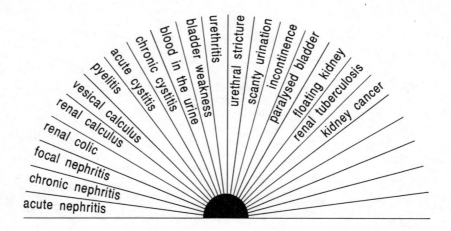

Figure 39. Kidney and bladder disease

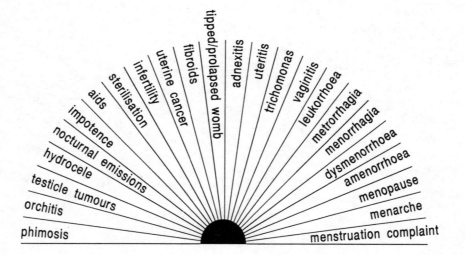

Figure 40. Disorders of the sex organs

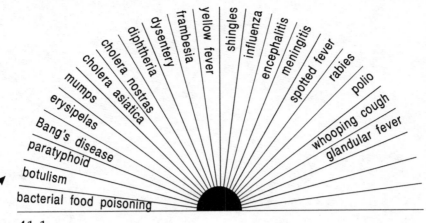

41.1

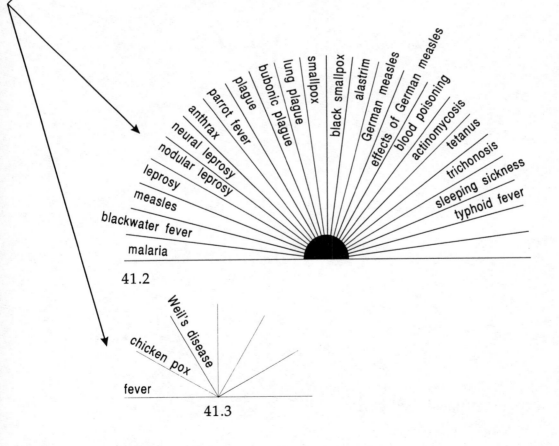

41.2

41.3

Figure 41. Infectious diseases

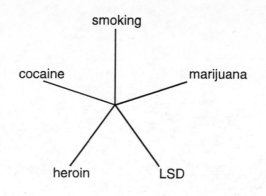

smoking

cocaine   marijuana

heroin   LSD

42.1

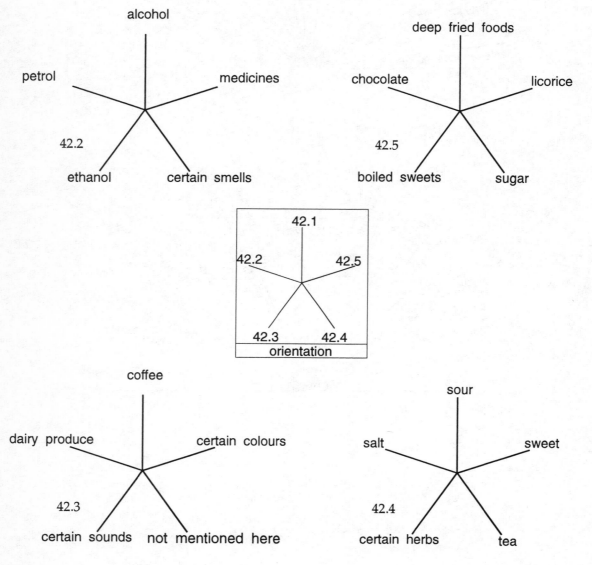

alcohol

petrol   medicines

42.2

ethanol   certain smells

deep fried foods

chocolate   licorice

42.5

boiled sweets   sugar

42.1

42.2   42.5

42.3   42.4

orientation

coffee

dairy produce   certain colours

42.3

certain sounds   not mentioned here

sour

salt   sweet

42.4

certain herbs   tea

Figure 42. Addictions (see also chapter 11 on 'Allergy')

6. Treatments

There are ailments and diseases that can be cured by the patient himself, but some need adequate medical care. Most of the therapies in use today are listed on the following page. The pendulum may help you to find the most appropriate therapy for a specific person. Once you have determined the best treatment, you can also use the pendulum to consult a list of addresses to find the best therapist. Some diagrams have been added here which may enable you to find out why a certain treatment or therapy has not got the desired effect.

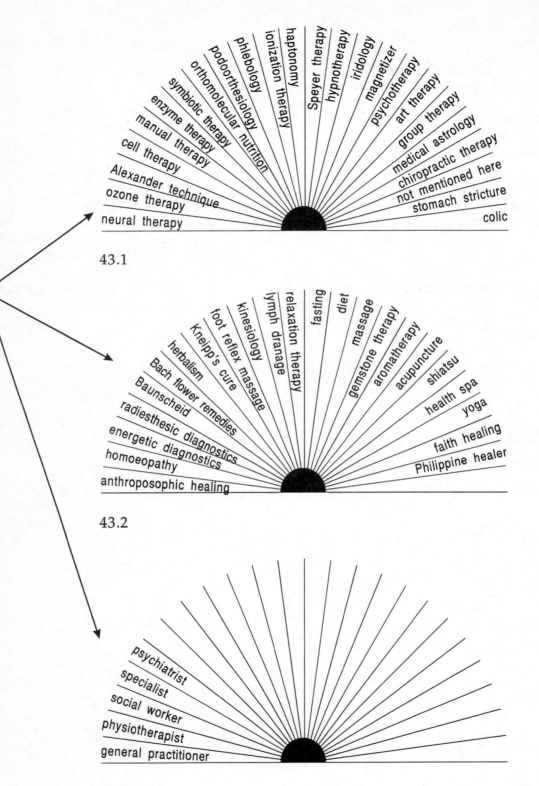

43.1

- neural therapy
- ozone therapy
- Alexander technique
- cell therapy
- manual therapy
- enzyme therapy
- symbiotic therapy
- orthomolecular nutrition
- podoorthesiology
- phlebology
- ionization therapy
- haptonomy
- Speyer therapy
- hypnotherapy
- iridology
- magnetizer
- psychotherapy
- art therapy
- group therapy
- medical astrology
- chiropractic therapy
- not mentioned here
- stomach stricture
- colic

43.2

- anthroposophic healing
- homoeopathy
- energetic diagnostics
- radiesthesic diagnostics
- Baunscheid
- Bach flower remedies
- herbalism
- Kneipp's cure
- foot reflex massage
- kinesiology
- lymph dranage
- relaxation therapy
- fasting
- diet
- massage
- gemstone therapy
- aromatherapy
- acupuncture
- shiatsu
- health spa
- yoga
- faith healing
- Philippine healer

43.3

- general practitioner
- physiotherapist
- social worker
- specialist
- psychiatrist

Figure 43. Therapies and healers

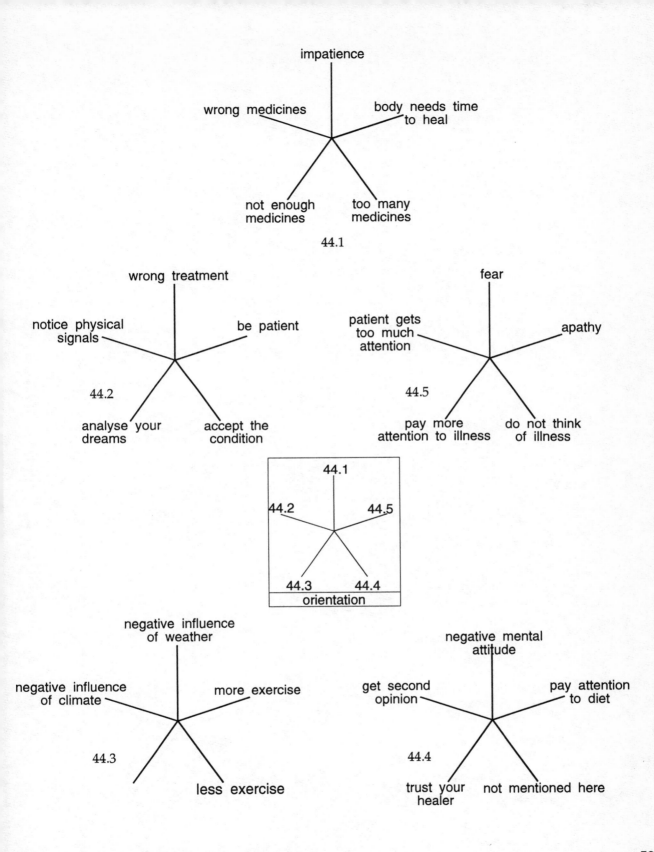

Figure 44. Causes of failing treatment
suggestions for improvement

59

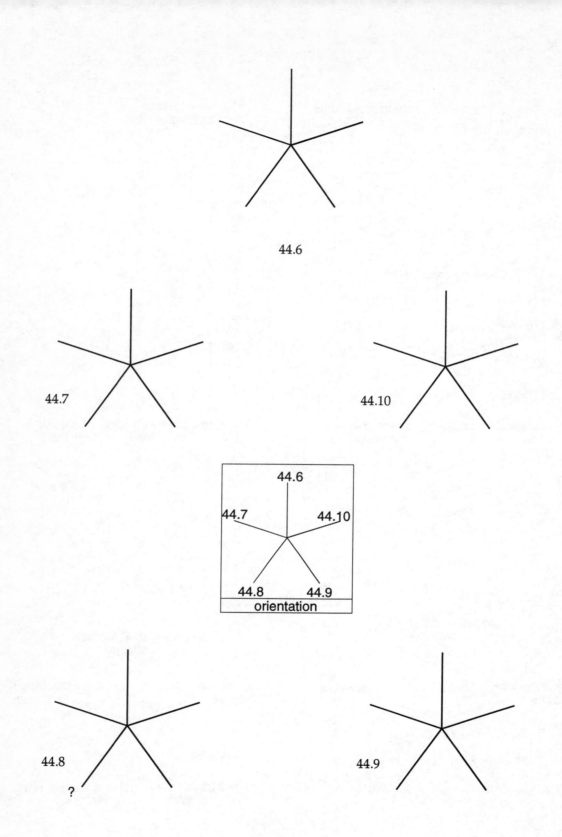

44.6

44.7

44.10

44.6

44.7

44.10

44.8

44.9

orientation

44.8

?

44.9

7. Vitamins and minerals

Nourishing food is often hard to find these days; in most cases food is sprayed with insecticides, or refined, coloured, tinned, or contains various unnecessary additives. Not surprisingly, deficiencies may occur. Furthermore, people who are just recovering from a serious illness - or suffer from stress over a long period of time - may need something extra.

With the help of the diagrams on the following pages you may determine which vitamins or minerals a person could use more of. You can also use the pendulum to determine the right dose. If there are several brands to choose from, you can even use the pendulum to find out which is best.

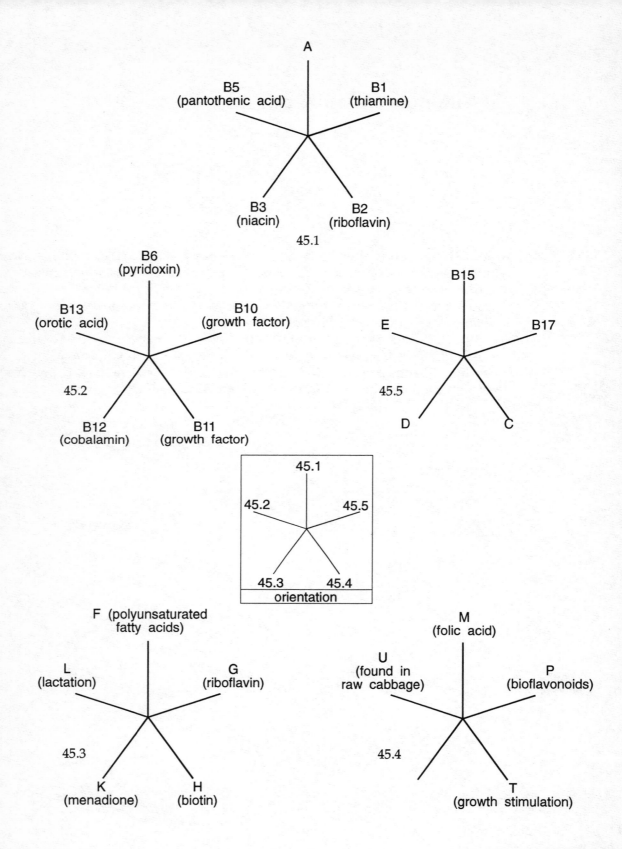

A

B5 (pantothenic acid)   B1 (thiamine)

B3 (niacin)   B2 (riboflavin)

45.1

B6 (pyridoxin)

B13 (orotic acid)   B10 (growth factor)

45.2

B12 (cobalamin)   B11 (growth factor)

B15

E   B17

45.5

D   C

45.1

45.2   45.5

45.3   45.4

orientation

F (polyunsaturated fatty acids)

L (lactation)   G (riboflavin)

45.3

K (menadione)   H (biotin)

M (folic acid)

U (found in raw cabbage)   P (bioflavonoids)

45.4

T (growth stimulation)

Figure 45. Vitamins

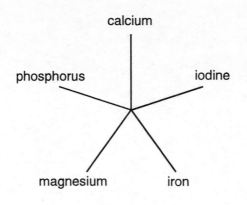
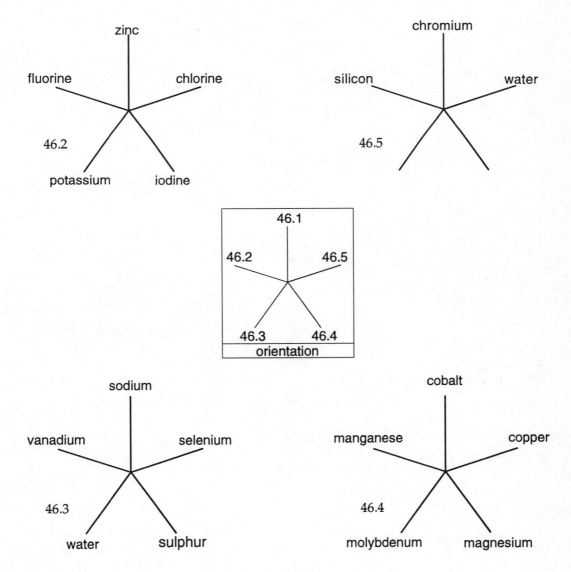

Figure 46. Minerals

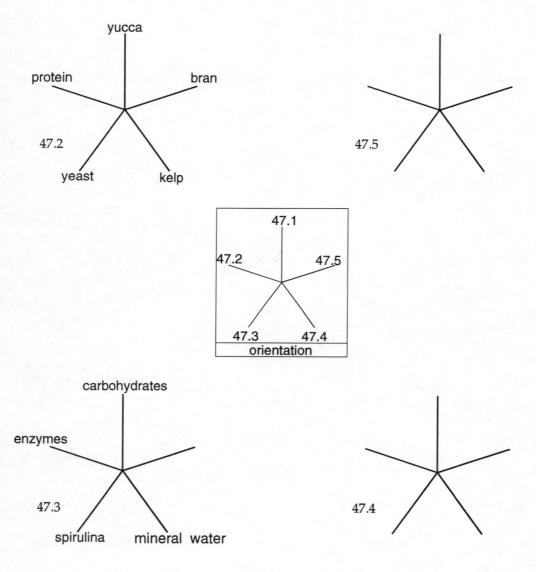

acidophilus

chlorophyll   ginseng

garlic   alfalfa

47.1

yucca

protein   bran

47.2

yeast   kelp

47.5

47.1

47.2   47.5

47.3   47.4

orientation

carbohydrates

enzymes

47.3

spirulina   mineral water

47.4

64

Figure 47. Other important substances

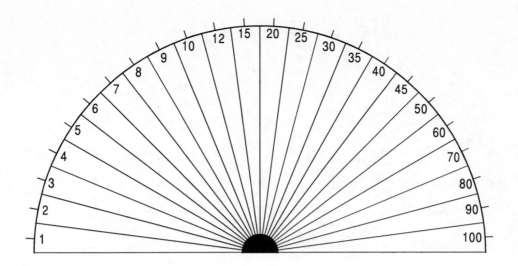

Figure 48. Dosage

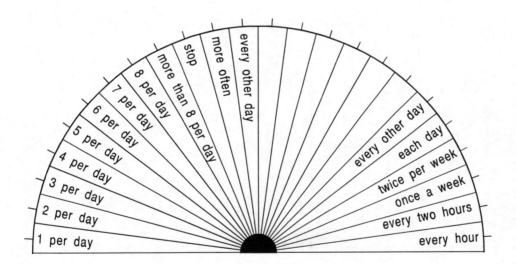

Figure 49. Use level

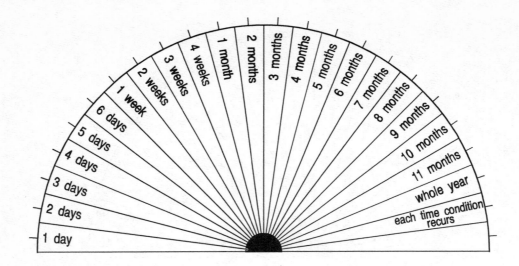

Figure 50. Period

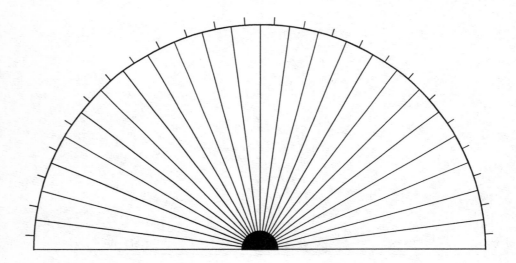

66

8. Homoeopathy

Homoeopathy is a wonderful method of treatment. It is now used more and more. Unfortunately, it is rather time-consuming to find the right medicine. Experienced classical homoeopaths spend a lot of time on this. It is especially complicated by the fact that symptoms may not always manifest themselves equally clearly in patients with a similar disease.

The following diagrams will prove to be helpful in this respect. There are some diagrams that help to determine the potency or dose. In order to ascertain how often and how long a medicine needs to be taken, please refer to figures 48-50 from the previous chapter. The diagrams on page 69 refer to one of the figures found in the last pages of this chapter, which lists several homoeopathic medicines. You can use the pendulum to determine the right medicine per page.

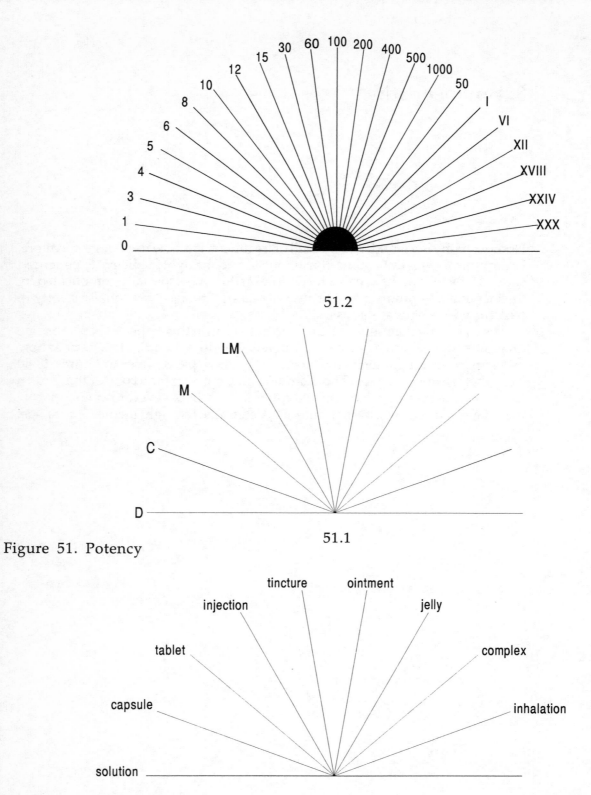

51.2

51.1

Figure 51. Potency

Figure 52. Type of medicine

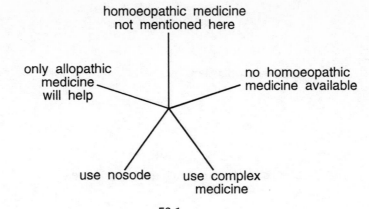

homoeopathic medicine
not mentioned here

only allopathic
medicine
will help

no homoeopathic
medicine available

use nosode   use complex
medicine

53.1

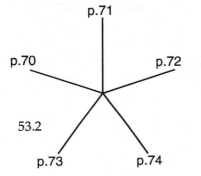

p.71

p.70   p.72

53.2

p.73   p.74

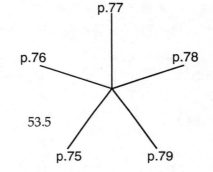

p.77

p.76   p.78

53.5

p.75   p.79

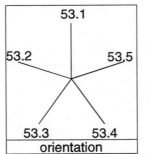

53.1

53.2   53.5

53.3   53.4
orientation

p.82

p.81   p.83

53.3

p.80   p.87

p.85

p.86   p.88

53.4

p.87

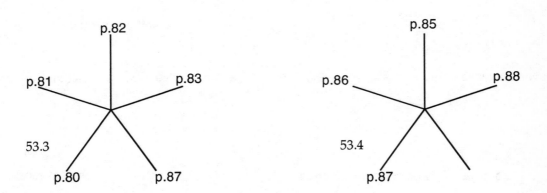

Figure 53. Number of page listing the right medicine

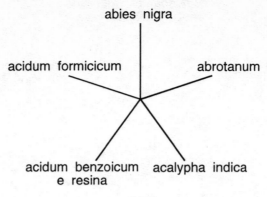

abies nigra

acidum formicicum

abrotanum

acidum benzoicum
e resina

acalypha indica

54.1

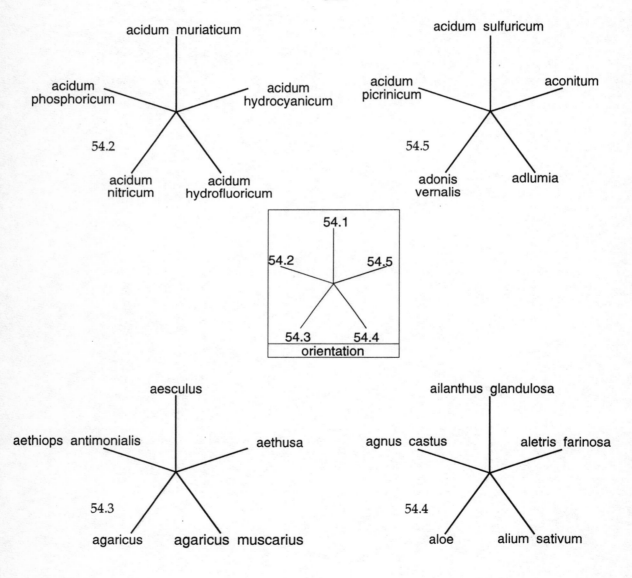

acidum muriaticum

acidum
phosphoricum

acidum
hydrocyanicum

54.2

acidum
nitricum

acidum
hydrofluoricum

acidum sulfuricum

acidum
picrinicum

aconitum

54.5

adonis
vernalis

adlumia

54.1

54.2

54.5

54.3

54.4

orientation

aesculus

aethiops antimonialis

aethusa

54.3

agaricus

agaricus muscarius

ailanthus glandulosa

agnus castus

aletris farinosa

54.4

aloe

alium sativum

Figure 54. Homoeopathic medicines

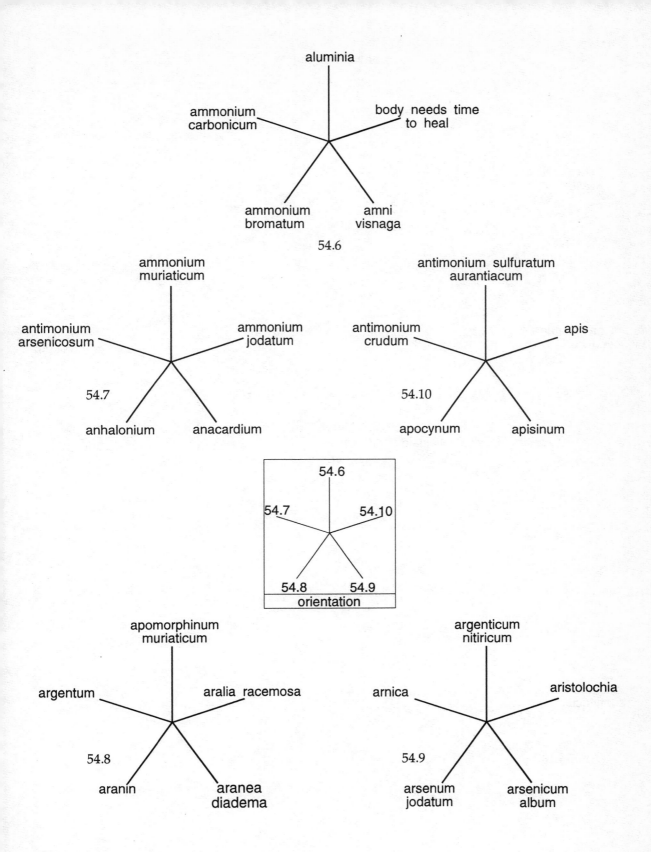

aluminia

ammonium
carbonicum

body needs time
to heal

ammonium
bromatum

amni
visnaga

54.6

ammonium
muriaticum

antimonium
arsenicosum

ammonium
jodatum

54.7

anhalonium

anacardium

antimonium sulfuratum
aurantiacum

antimonium
crudum

apis

54.10

apocynum

apisinum

54.6

54.7

54.10

54.8

54.9

orientation

apomorphinum
muriaticum

argentum

aralia racemosa

54.8

aranin

aranea
diadema

argenticum
nitiricum

arnica

aristolochia

54.9

arsenum
jodatum

arsenicum
album

71

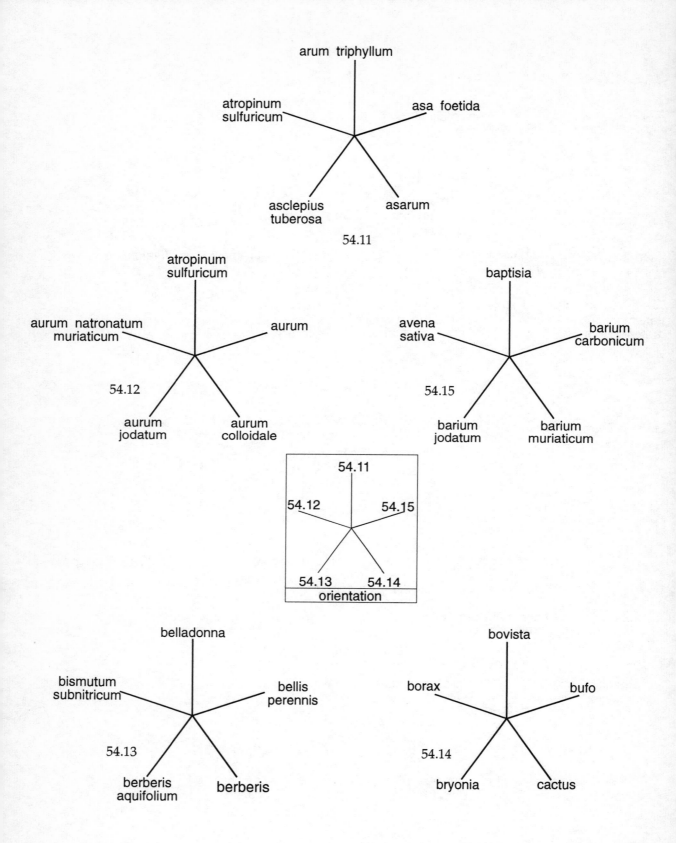

arum triphyllum

atropinum
sulfuricum

asa foetida

asclepius
tuberosa

asarum

54.11

atropinum
sulfuricum

aurum natronatum
muriaticum

aurum

54.12

aurum
jodatum

aurum
colloidale

baptisia

avena
sativa

barium
carbonicum

54.15

barium
jodatum

barium
muriaticum

54.11

54.12   54.15

54.13   54.14
orientation

belladonna

bismutum
subnitricum

bellis
perennis

54.13

berberis
aquifolium

berberis

bovista

borax

bufo

54.14

bryonia

cactus

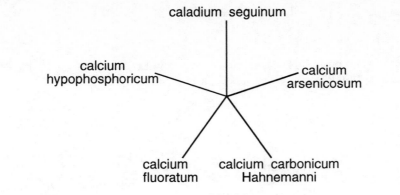

caladium seguinum

calcium hypophosphoricum

calcium arsenicosum

calcium fluoratum

calcium carbonicum Hahnemanni

54.16

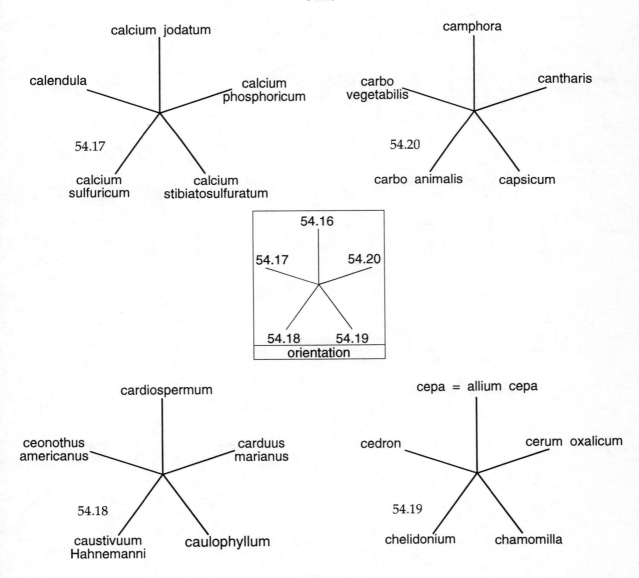

calcium jodatum

calendula

calcium phosphoricum

54.17

calcium sulfuricum

calcium stibiatosulfuratum

camphora

carbo vegetabilis

cantharis

54.20

carbo animalis

capsicum

54.16

54.17

54.20

54.18

54.19

orientation

cardiospermum

ceonothus americanus

carduus marianus

54.18

caustivuum Hahnemanni

caulophyllum

cepa = allium cepa

cedron

cerum oxalicum

54.19

chelidonium

chamomilla

73

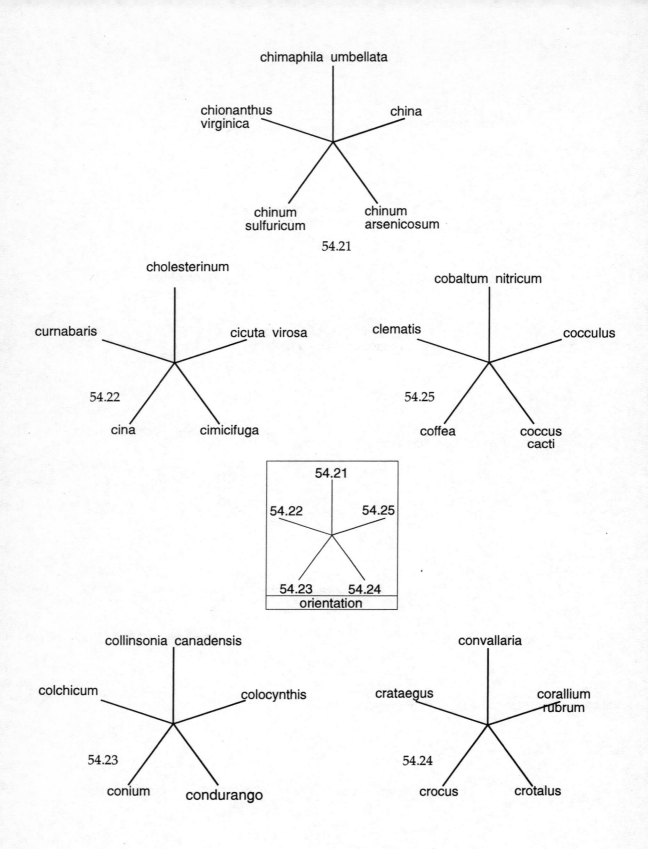

chimaphila umbellata

chionanthus virginica

china

chinum sulfuricum

chinum arsenicosum

54.21

cholesterinum

curnabaris

cicuta virosa

54.22

cina

cimicifuga

cobaltum nitricum

clematis

cocculus

54.25

coffea

coccus cacti

54.21

54.22

54.25

54.23

54.24

orientation

collinsonia canadensis

colchicum

colocynthis

54.23

conium

condurango

convallaria

crataegus

corallium rubrum

54.24

crocus

crotalus

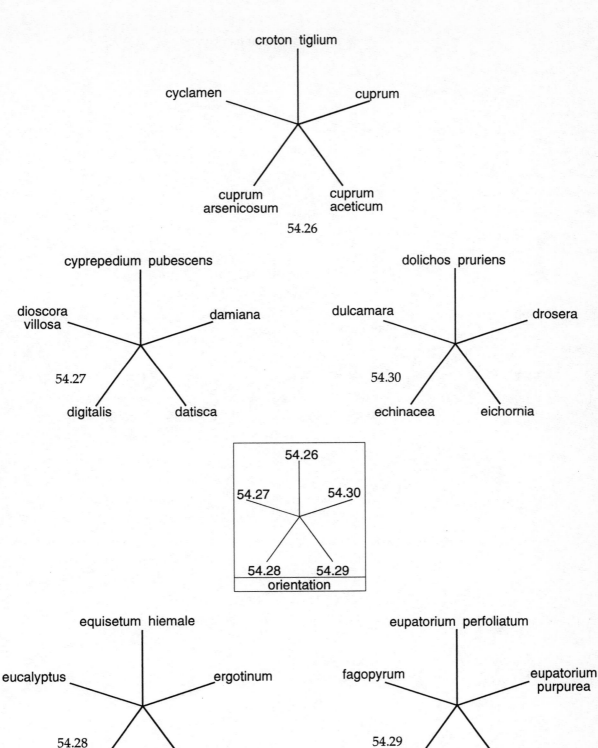

croton tiglium

cyclamen   cuprum

cuprum   cuprum
arsenicosum   aceticum

54.26

cyprepedium pubescens

dioscora
villosa   damiana

54.27

digitalis   datisca

dolichos pruriens

dulcamara   drosera

54.30

echinacea   eichornia

54.26

54.27   54.30

54.28   54.29

orientation

equisetum hiemale

eucalyptus   ergotinum

54.28

espeletia   erigeron
                 canadensis

eupatorium perfoliatum

fagopyrum   eupatorium
                 purpurea

54.29

euphrasia   euphorbium

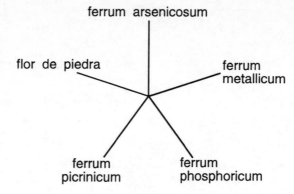

ferrum arsenicosum

flor de piedra   ferrum metallicum

ferrum picrinicum   ferrum phosphoricum

54.31

formica rufa

ginseng   fucus vesiculosus

54.32

gelsenium   galphimia

glonoium

guajacum   gnaphalium

54.35

grindelia   graphites

54.31

54.32   54.35

54.33   54.34

orientation

hamamelis

heleborus niger   haplopappus

54.33

hekla lava   harpagophytum

helonias dioica

hyoscyamus   heparsulfuris

54.34

hydrocotyle asiatica   hydrastis

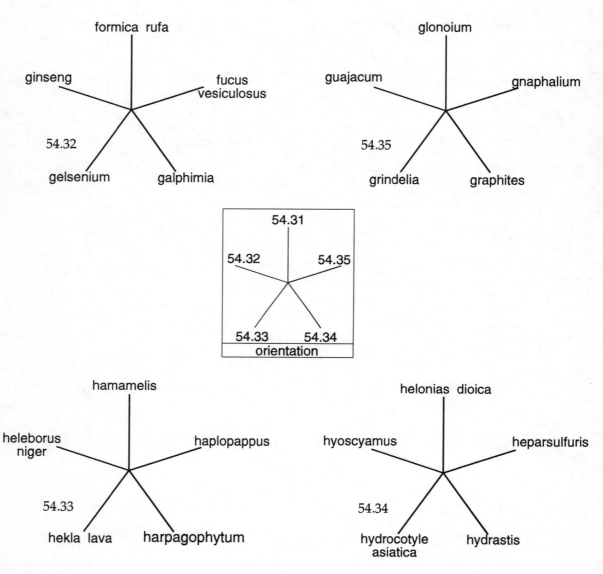

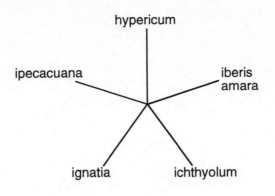

hypericum

ipecacuana

iberis
amara

ignatia

ichthyolum

54.36

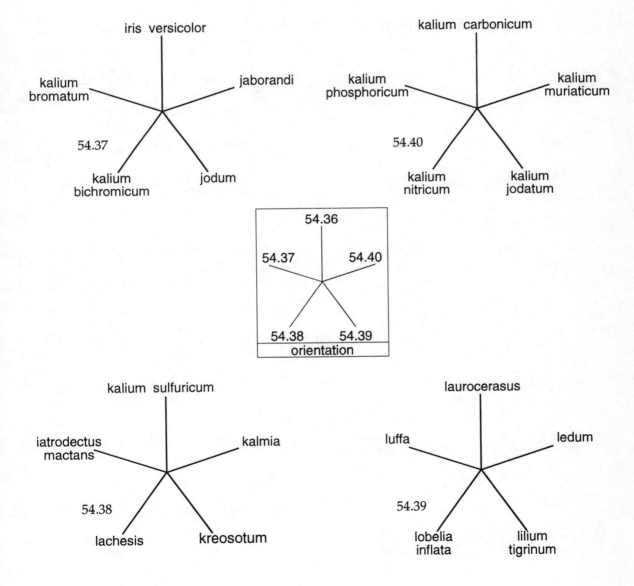

iris versicolor

kalium
bromatum

jaborandi

54.37

kalium
bichromicum

jodum

kalium carbonicum

kalium
phosphoricum

kalium
muriaticum

54.40

kalium
nitricum

kalium
jodatum

54.36

54.37

54.40

54.38

54.39

orientation

kalium sulfuricum

iatrodectus
mactans

kalmia

54.38

lachesis

kreosotum

laurocerasus

luffa

ledum

54.39

lobelia
inflata

lilium
tigrinum

77

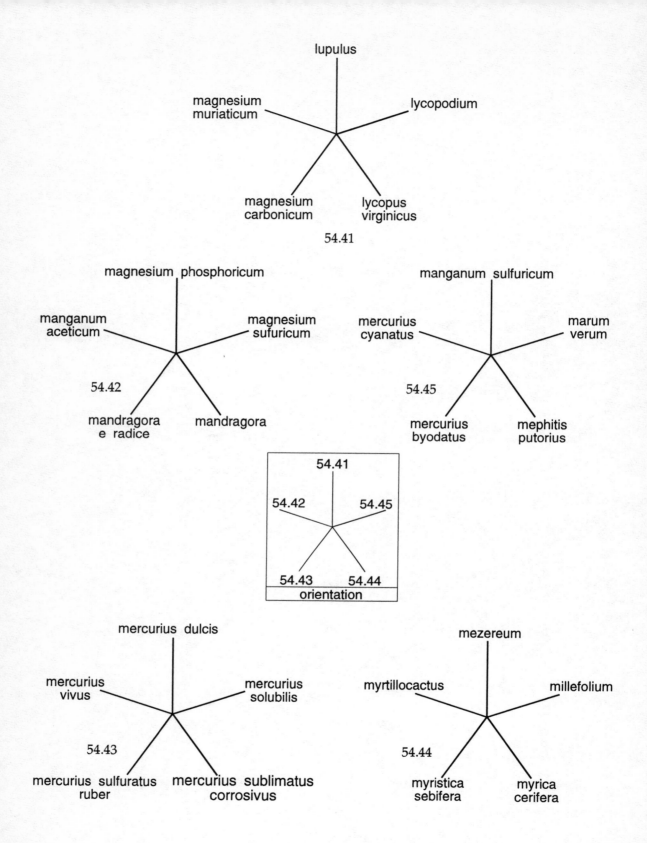

lupulus

magnesium
muriaticum

lycopodium

magnesium
carbonicum

lycopus
virginicus

54.41

magnesium phosphoricum

manganum
aceticum

magnesium
sufuricum

54.42

mandragora
e radice

mandragora

manganum sulfuricum

mercurius
cyanatus

marum
verum

54.45

mercurius
byodatus

mephitis
putorius

54.41

54.42

54.45

54.43

54.44

orientation

mercurius dulcis

mercurius
vivus

mercurius
solubilis

54.43

mercurius sulfuratus
ruber

mercurius sublimatus
corrosivus

mezereum

myrtillocactus

millefolium

54.44

myristica
sebifera

myrica
cerifera

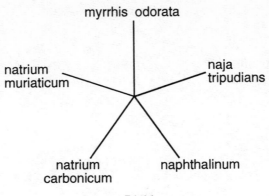

myrrhis odorata

natrium muriaticum

naja tripudians

natrium carbonicum

naphthalinum

54.46

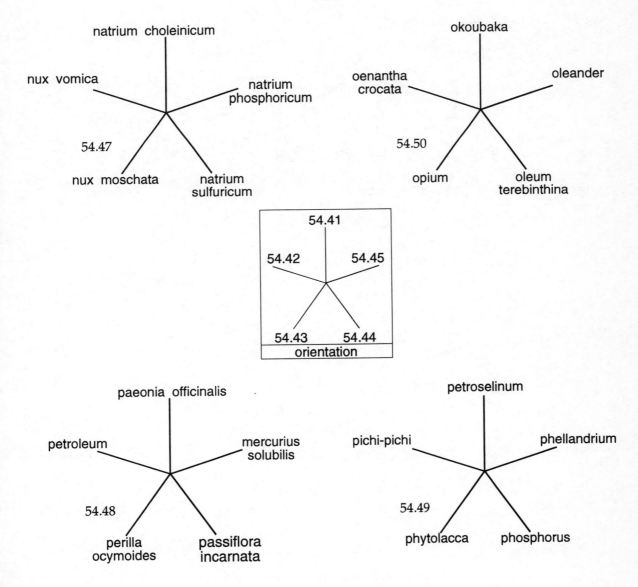

natrium choleinicum

nux vomica

natrium phosphoricum

54.47

nux moschata

natrium sulfuricum

okoubaka

oenantha crocata

oleander

54.50

opium

oleum terebinthina

54.41

54.42

54.45

54.43

54.44

orientation

paeonia officinalis

petroleum

mercurius solubilis

54.48

perilla ocymoides

passiflora incarnata

petroselinum

pichi-pichi

phellandrium

54.49

phytolacca

phosphorus

79

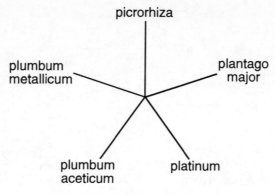

picrorhiza

plumbum
metallicum

plantago
major

plumbum
aceticum

platinum

54.51

podophyllum

pulsatilla

populus
tremuloides

54.52

prunus
spinosa

potentilla
anserina

pyrogenium

rheum

quassia

54.55

rauwolfia
serpentina

ranunculus
bulbosa

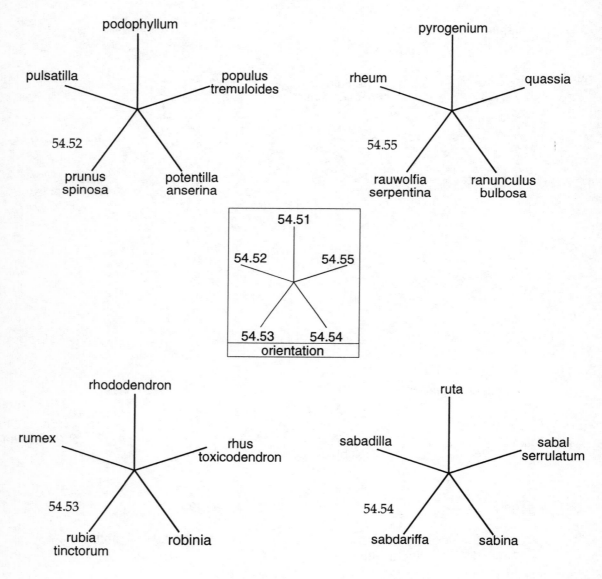

54.51

54.52

54.55

54.53

54.54

orientation

rhododendron

rumex

rhus
toxicodendron

54.53

rubia
tinctorum

robinia

ruta

sabadilla

sabal
serrulatum

54.54

sabdariffa

sabina

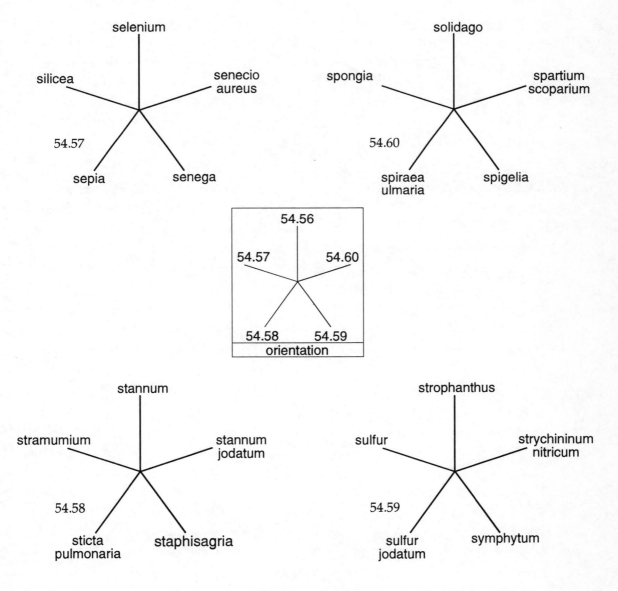

sambucus nigra

secale cornutum

sanguinaria

scilla

sarsaparilla

54.56

selenium

silicea

senecio aureus

54.57

sepia

senega

solidago

spongia

spartium scoparium

54.60

spiraea ulmaria

spigelia

	54.56	
54.57		54.60
54.58		54.59
	orientation	

stannum

stramumium

stannum jodatum

54.58

sticta pulmonaria

staphisagria

strophanthus

sulfur

strychininum nitricum

54.59

sulfur jodatum

symphytum

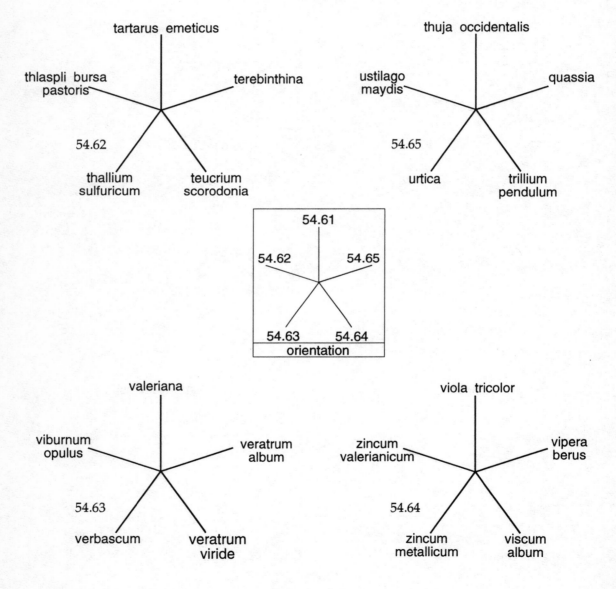

symphytum ad usum externum

taraxacum

syzygium
janbolanum

tarantula   tabacum

54.61

tartarus emeticus

thlaspli bursa
pastoris

terebinthina

54.62

thallium
sulfuricum

teucrium
scorodonia

thuja occidentalis

ustilago
maydis

quassia

54.65

urtica

trillium
pendulum

54.61

54.62   54.65

54.63   54.64

orientation

valeriana

viburnum
opulus

veratrum
album

54.63

verbascum

veratrum
viride

viola tricolor

zincum
valerianicum

vipera
berus

54.64

zincum
metallicum

viscum
album

82

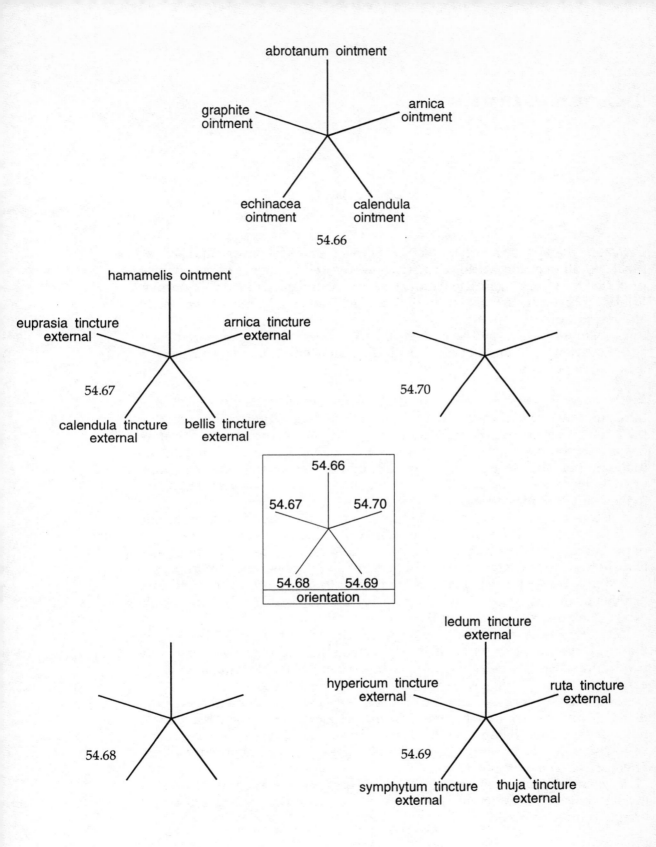

abrotanum ointment

graphite
ointment

arnica
ointment

echinacea
ointment

calendula
ointment

54.66

hamamelis ointment

euprasia tincture
external

arnica tincture
external

54.67

calendula tincture
external

bellis tincture
external

54.70

54.66

54.67   54.70

54.68   54.69

orientation

54.68

ledum tincture
external

hypericum tincture
external

ruta tincture
external

54.69

symphytum tincture
external

thuja tincture
external

83

9. Gemstone therapy

Gemstone therapy is an ancient method of treatment, which was - and still is - practised all over the world. Gemstones were used to make amulets, which protected the wearer against evil influences, or accompanied him on his journey. Priests and clergymen wore -and still wear - gemstones to increase their insight and inspiration.

Gemstones were also used in medicine. Apothecaries sold powdered gems against various ailments and diseases. Today, gemstone therapy again enjoys a growing popularity.

The effect of gems can be explained as follows. A gemstone is really a living thing; it, however, grows so slowly - one millimetre in a thousand years - that this is not visible to us. All living creatures have an aura. Every gemstone therefore has an aura specific to its kind. When a person is ill, his or her aura does not radiate in the proper way. A gemstone may make up this deficit. The malfunctioning organ is 'recharged'. To this end, it is important to wear the gem on the skin close to the malfunctioning organ.

The interaction between gem and bearer can be very strong. Sometimes the symptoms will first become worse before improvement sets in. It is very important to purify the gemstone regularly. After a certain period of time it has absorbed so much energy from the ill person that it cannot function properly.

A simple way of purifying or cleaning a gem is to immerse it in running water. Another method is to bury it in sand (ordinary sand, not potting soil). Some people expose their gems to the light of the full moon for a night.

The gem needs to cleaned once every fortnight, at the beginning of a gemstone therapy sometimes even more often. This can be checked simply by asking the pendulum: Does this gemstone need to be cleaned? After cleaning it, you can ask: Is this stone completely clean?

After a certain period of time the gemstone may have 'given' so much energy that it loses its lustre or colour, even cracks may appear. In such case, it is advisable to use a new stone. There are positive, yang stones. These are red, orange, yellow, yellowish green, terracotta or gold-coloured. They have an invigorating effect. There are also negative yin stones.

These are blue, bluish green, purple, brown, grey or silver coloured. They reject evil and poison. Furthermore, there are neutral stones, which are green, beige, brandy or liver coloured.

Gemstones of the Zodiac

Capricorn the Goat (22 Dec.-20 Jan.): smoky quartz, onyx, jet
Aquarius the Waterbearer (21 Jan.-18 Feb.): turquoise, amazonite, malachite
Pisces the Fish (19 Feb.-20 Mar.):amethyst, moonstone, opal, aquamarine
Aries the Ram (21 Mar.-20 Apr.): red jasper, carnelian, ruby
Taurus the Bull (21 Apr.-20 May): blue sapphire, rose quartz, lapis
Gemini the Twins (21 May-20 June): topaz, citrine, tiger's eye, rock crystal, aquamarine, chalcedony
Cancer the Crab (21 June-20 July): emerald, white chalcedony, aventurine, chrysoprase, peridote
Leo the Lion (21 July-22 Aug.): garnet, rock crystal, diamond, peridote, onyx
Virgo the Virgin (23 Aug.-22 Sept.): carnelian, yellow agate, layered onyx
Libra the Scales (23 Sept.-22 Oct.): aventurine, jade, nephrite, emerald
Scorpio the Scorpion (23 Oct.-22 Nov.): hematite, garnet, red tourmaline
Sagittarius the Archer (22 Nov.-21 Dec.): topaz, chalcedony

Gemstones used in gemstone therapy

Agate:	calming, strengthens the heart, against pain, deafness, poisoning, homesickness, buzzing in the ears, and fever
Agate (Botswana):	eye troubles
Amazonite:	eczema, boils
Aquamarine:	zest for life, against allergy, thyroid disorders, struma, cold, throat affections
Aventurine:	calming, improves self-control, against skin diseases such as acne, eczema, nettle rash, and psoriasis, also against nosebleed
Amethyst:	inner calm, stability, depurant effect, against migraine and other headaches, insomnia, swellings, and infections of wrist and knee
Amber:	against fear, overtiredness, and asthma
Rock crystal:	conquers fear, against backaches, disturbance of equilibrium, carsickness, seasickness, abdominal spasms, diarrhoea, and menstrual disorders
Beryl:	good for throat and eyes, against hepatitis
Red coral:	against high blood pressure
Chalcedony:	against depressive moods, bleeding, chillblained hands, splits in lips/hands
Carnelian:	spirit of enterprise, against cramps, rheumatism, fever, infections, diabetes, fear of exams, nightmares, melancholy, neuralgia
Citrine:	nerve-strengthening, invigorating
Chrysocolla:	strengthens and harmonizes the solar plexus, against arthritis and arthrosis

Chrysoprase:	for sea voyages, increases the longing for insight and higher consciousness, improves eyesight
Garnet:	improves willpower and intuition, against depression, melancholy, nervous exhaustion, strengthens the heart
Heliotrope:	gives comfort and courage under adverse conditions, against haemorrhoids, stones in the bladder, nosebleed
Hematite:	gives courage, improves sleep, strengthens the heart
Jade:	stone of the five great virtues: modesty, charity, courage, justice, and wisdom; against bed-wetting, strenghtens kidney and bladder
Jasper:	against epilepsy, stomach disorders, bladder disorders, morning sickness, gall, liver, and kidney troubles
Coral:	improves blood, loses colour when wearer is anaemic
Labradorite:	strong beneficial effect on many organs, suits anyone, increases intuition
Lapis lazuli:	improves sleep, against epilepsy and strokes, beneficial to heart, spleen, blood, and skin, increases love and self-confidence, stone of friendship
Moonstone:	against misfortune, infertility, poisoning, menstrual disorders, improves inspiration and provides protection on journeys
Magnetite:	against neuralgia, rheumatism, fractures, cramp in the calf
Malachite:	against asthma, irregular menstruation, aching joints, rheumatism, beneficial effect on multiple sclerosis, Parkinson's disease, varicose veins, and boils
Nephrite:	all diseases of the lumbar/renal region
Olivine (peridote):	gives eloquence, improves heart and eyes, beneficial effect on slipped disc, depression, instability, hair loss, menstrual disorders
Obsidian:	beneficial effect on eye trouble, cataract, improves eyesight
Onyx:	against shortness of breath, running eyes, calcium deficiency, deafness, improves balance, and growth of hair and nails
Pyrites:	purifies the body and aura, improves breathing, against throat infections
Rhodochrosiet:	calming, beneficial effect on Parkinson's disease, fear, and confusion
Rhodonite:	improves the power of reason, has beneficial effect on multiple sclerosis
Ruby:	prevents miscarriage, dispels melancholy and fear
Smoky quartz:	calming, helps to give up smoking, improves hearing and the ability to concentrate

Rose quartz:	stimulates thymus gland, against stomach cramps, abdominal disorders, dizziness, and homesickness, improves love, art, creative thinking, and imagination
Rutilated quartz:	against bronchitis and asthma, beneficial to bronchial tubes, warming effect
Sapphire:	for sincerity and fidelity, beneficial to heart and eyes
Emerald:	against epilepsy, improves memory, beneficial to eyes
Sodalite:	diligence, against high blood pressure and nervous exhaustion
Tourmaline:	against falling, stumbling, disturbances of equilibrium, dizziness, hyperventilation, stuttering, neck troubles, effects of radiation, improves ability to concentrate, self-confidence, and vitality
Topaz:	against varicose veins, thrombosis, insomnia, exhaustion, stimulates pancreas, improves liver and organ of taste
Turquoise:	improves meditation, discolours when wearer is in danger
Tiger's eye:	against cold, cramps, protects and has warming effect
Unakite (epidote):	against infections, liver and stomach troubles, improves willpower
Falcon's eye:	against shortness of breath, beneficial to bronchial tubes, improves clear thinking

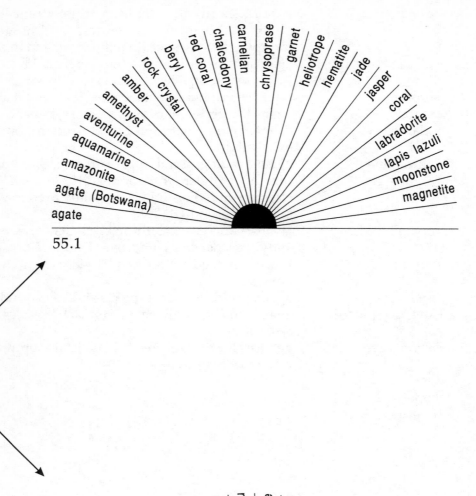

55.1

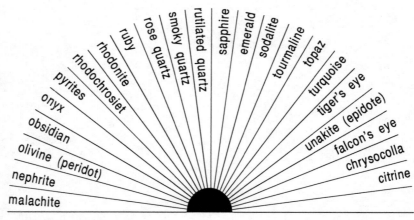

55.2

Figure 55. Gemstones

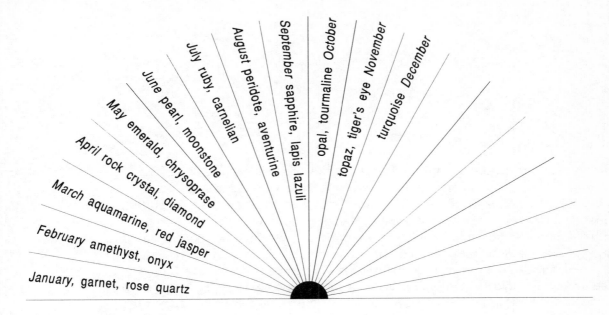

Figure 56. Birth-stones

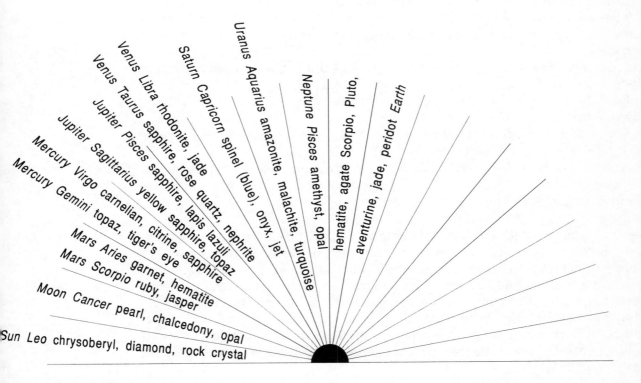

Figure 57. Stones of the planets

10. Radiation

Radiation is omnipresent in nature. Certain types of radiation are beneficial to man, other types are less beneficial or even constitute a serious threat. Sunlight is radiation, too. We like basking in the sun. Yet, if this radiation is concentrated, for instance by means of a burning glass, then it is not wise to focus it on your skin. This shows that harmless radiation becomes so powerful when concentrated that it can cause serious damage. A great amount of radiation is released in space and the earth itself as a result of various cosmic processes. If this radiation is concentrated, it will cause problems. What kinds of problems it will cause are, on the one hand, determined by our own fitness and (over)sensitivity, and on the other hand, by the type of radiation and its intensity.

The following types of radiation and their effects can be distinguished:

Delta rays: from 0.5 up to 4 Hz - sleep, infirmity;

Theta rays: from 4 up to 8 Hz - deteriorating state of mind, feelings of superficiality, uninterestedness, negligence, overtiredness, emptiness, anxiety, worry;

Alpha rays: from 8 up to 13 Hz - you feel relaxed, passive, calm, tolerant, happy, glad;

Beta rays I: 14 up to 18/20 Hz - lethargic attitude changes into activity with the increase from 8 to 13 Hz or higher,

Intellectual or basic attitude more apparent, as in logical thinking, poetic explorations, searching, impatience, doubt;

Beta rays II: from 18/20 up to 35 Hz - even stronger effects, aggression;

Gamma rays: from 35 up to 90 Hz - aggression, very unhappy depressive state of mind, activates carcinogenic processes.

Generally, the magnetic field is distorted in such a way that loss of energy occurs. Also, the air becomes ionized, i.e. the composition of the air is changed so that there is a deficiency in negative ions, which contain oxygen. This radiation often feels like a 'cold draught'. Radiation in an over-insulated room may cause an increasing and lasting shortage of active oxygen. It is not surprising that someone who stays in such a room - particularly at night - will become seriously ill eventually. Radiation fluctuates in strength. There is an increase at night, during a new and full moon, and in low pressure areas. The term 'lunacy' can be associated with this.

When there is not enough active oxygen in a room, i.e. when there is a shortage of negative ions, the bad effects can be countered by using an ionizer. The best thing to do is not to stay in such rooms. A shortage of negative ions may also occur in air-conditioned buildings. Again, in such a case ionizers could be useful. Sometimes the lack of active oxygen may have a positive effect, as in the church of Wiewerd, a Frisian village, where they dug up mortal remains that were still intact after many years.

It is beyond the scope of this book to discuss the influences of various radiation sources such as power supplies, high-voltage cables, synthetic covering, underfloor heating, electric blankets, etc. The most important thing is to determine whether there is any radiation. Seeing the possible dangers of radiation, it is wise to get help from an expert.

With the help of the following diagrams you can determine the type of radiation you are dealing with, its intensity, the exact spot it comes from, and the effects on you and others.

In figure 60 three types of people are distinguished: a C-,W-, and M-type. A C-type of person is someone who reacts badly when the weather gets colder. A W-type is someone who reacts badly when the weather becomes warmer. The M-type of person reacts to both changes, but less fiercely.

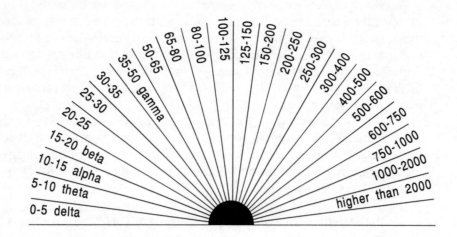

Figure 58. Radiation frequencies (Hz)

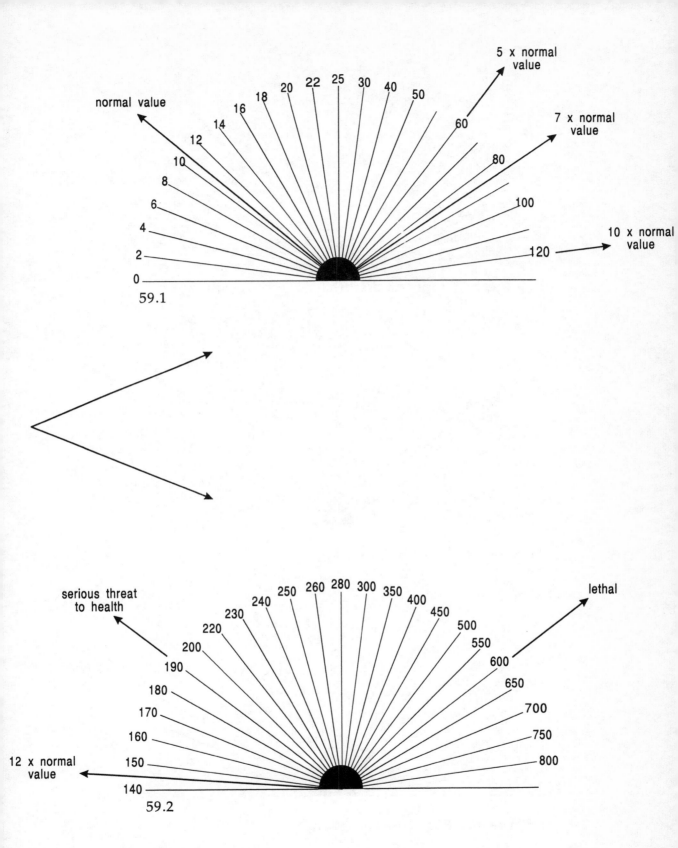

Figure 59. Radioactivity

93

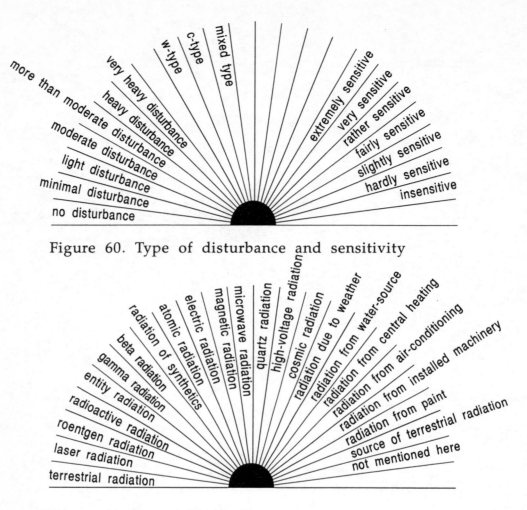

Figure 60. Type of disturbance and sensitivity

Figure 61. Type of radiation

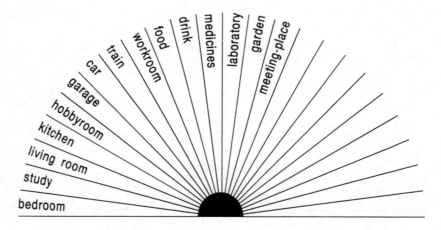

Figure 62. Focus of radiation

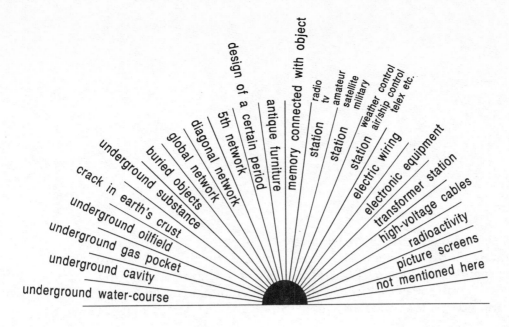

Figure 63. Radiation source

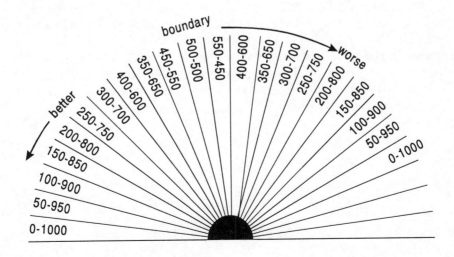

Figure 64. Positive and negative ions per 1000 ions

11. Allergy

Allergy is a disorder from which an increasing number of people seem to suffer. The most well-known form of allergy is hay fever. Particularly on a nice, sunny day, when lots of plants are in bloom, people with allergies develop the following symptoms: sneezing, running eyes, blocked nose, swollen mucous membranes of eyes, mouth, and nose, often headache, a sore throat, and a general feeling of not being well.

Some people react allergically and develop a rash after touching certain plants, working with certain chemicals, or after using cosmetics.

There are also less well-known manifestations of allergy, such as persistent eczema, chronic fatigue, abdominal cramps, and a swollen stomach in the case of food allergy, extreme changes of mood, itching, etc.

The diagrams given here contain some of the substances that can cause allergy. The list of substances is by no means complete. For food allergy, unfamiliar foodstuffs and complex products are not listed. Cosmetics and medicines are not mentioned by name. If you make a list of all the products you use, you can draw up your own diagrams. This also applies to substances used in the house. Sorting out the substances that are used professionally is a personal matter. An office clerk will come into contact with different substances from someone working in an asbestos plant, a car mechanic, farmer, or cook. You can use the diagrams that are left blank to fill in those substances that you are possibly allergic to. Of course you can also use the pendulum to check things out on the spot. Unfortunately for those suffering from hay fever, there is no complete list of plants which cause allergy. The number of wild and cultivated plants does not allow this. The diagrams that are left blank can be used to fill in all the plants in your environment. A detailed drawing of the neighbourhood might also help in this case.

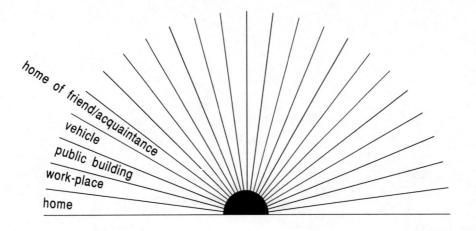

Figure 65.Location where allergy occurs

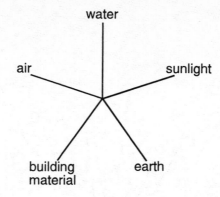

66.1

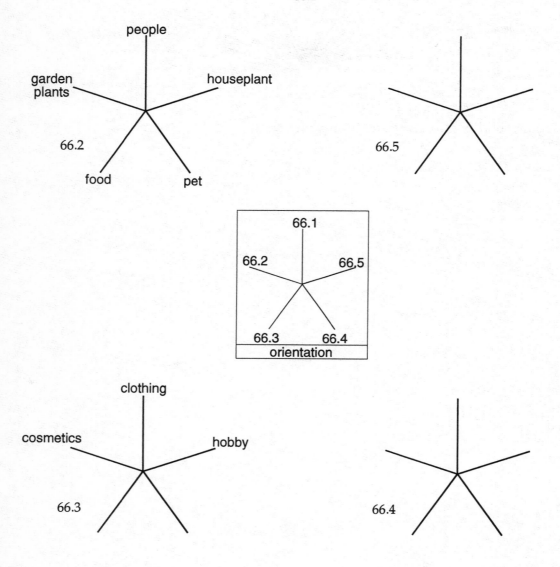

people

garden
plants

houseplant

66.2

food

pet

66.5

66.1

66.2

66,5

66.3

66.4

orientation

clothing

cosmetics

hobby

66.3

66.4

Figure 66. Some causes of allergy

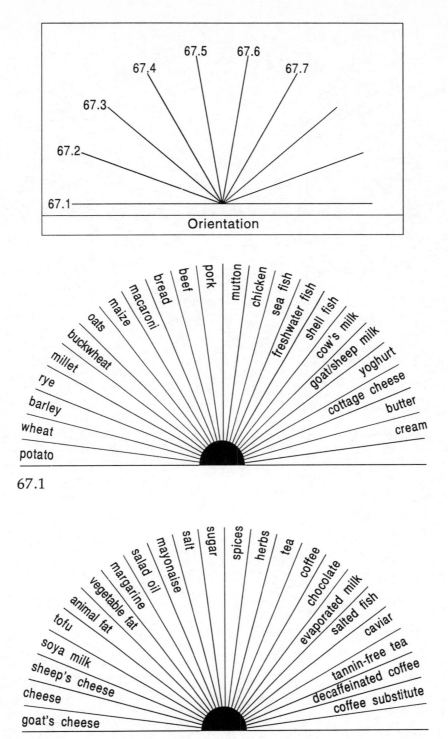

Orientation

67.1

4.8

Figure 67. Food causing allergy

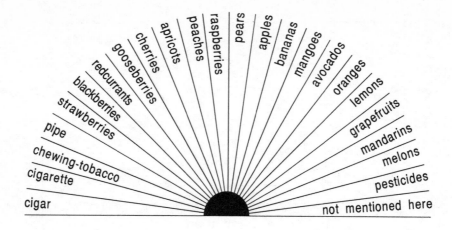

67.3

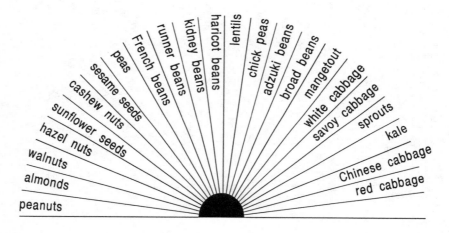

67.4

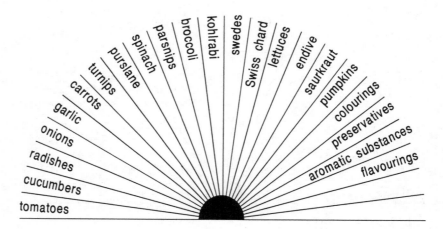

67.5

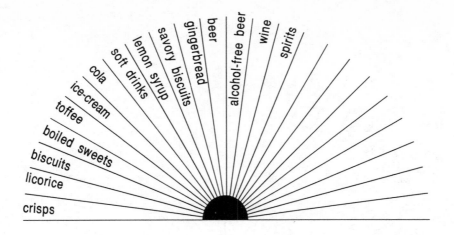

67.6

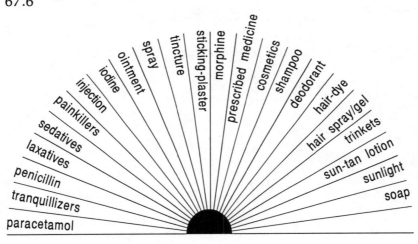

Figure 68. Medicines and cosmetics causing allergy

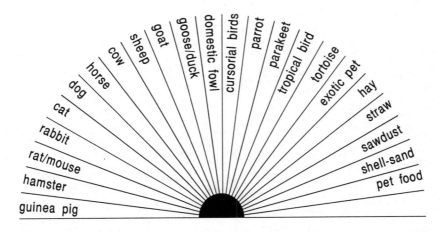

Figure 69. Pets causing allergy

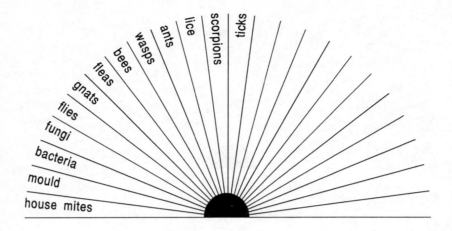

Figure 70. Parasites and vermin causing allergy

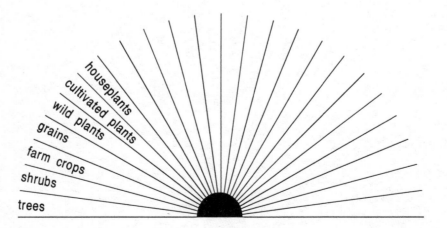

Figure 71. Plants causing allergy (hay fever)

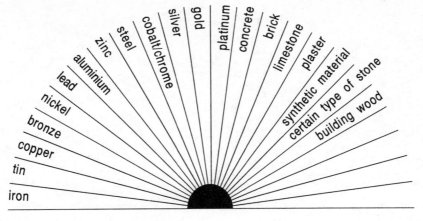

72.1

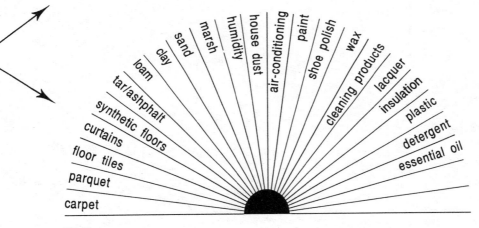

72.2

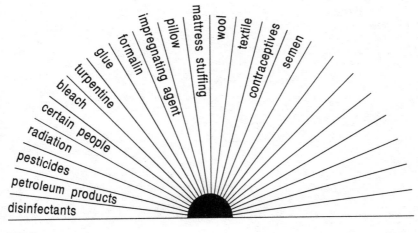

72.3

Figure 72. Other substances causing allergy   (home, work)

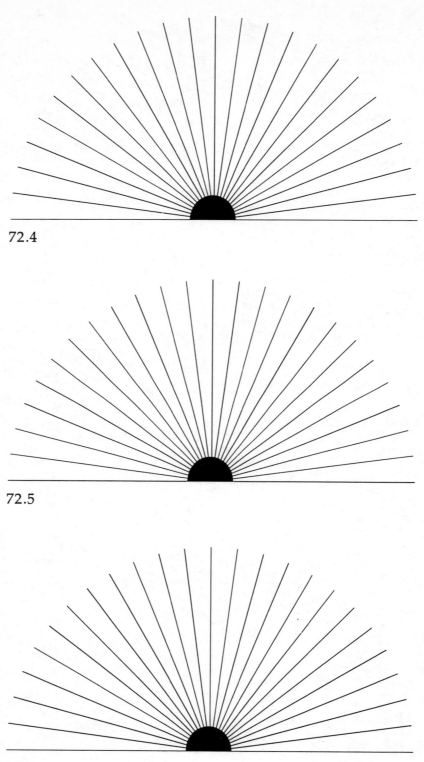

72.4

72.5

72.6

12. The aura

The aura should not be confused with the magnetic field that exists around all objects. The magnetic field around the human body is visible as a silver-coloured radiation reaching no further than a few centimetres. The aura extends much further and has a large pear, or onion-shaped radiation of transparent colours running into one another. Sometimes these colours are wonderful, sometimes they are dull. Often it is a mixture of colours in which one colour is dominant, depending on the person's mental make-up.

The chakras with their specific colours form this totality. For instance, someone of a stingy, avaricious and irascible nature will have a dominant root chakra and there will be much red in his aura. The dominant colour helps us to determine the physical and mental constitution of the person in question at a particular moment.

The colours have roughly the following meaning:

Blurry red	destructive impulses
Dark red	temperamentality
Crimson	passionate love
Brown red	hopelessness
Dull greenish red	stinginess
Carmine red	human affection
Infrared	psychic tendency to lose one's self
Bright red	physical love
Deep scarlet red	sensual love
Rose red	cheerfulness
Mussel red	creativity
Coral	immaturity, juvenility
Lively red	vitality
Orange	fading life, dying
Luminous orange	pride
Orange-yellow	ambition
Yellow	spirituality, creative love, clear thinking focused on higher reality
Mustard yellow	meaness, slyness
Gold	thoughts focused on higher reality
Green	adaptability, beginning of healing power
Grass-green	love for music

Jade green	wordly wisdom
Dark green	envy, jealousy
Fresh green	sympathy
Blue	trust in oneself
Luminous blue	confidence
Grey blue	superstition
Royal blue	loyalty
Pale blue	lack of depth
Indigo	egoism
Violet	intuition
Deep violet	religiosity
Purple	well-developed healing powers
Brown	narrow-mindedness, greed
Reddish brown	greed, covetous
Grey	fear
Black	enveloped by mystery, death
White	perfection

With the corresponding diagrams you can determine the dominant colour.

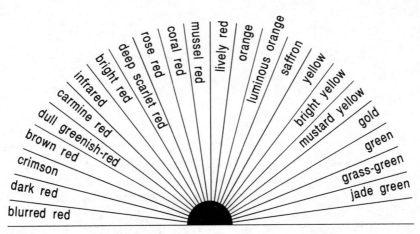

73.1

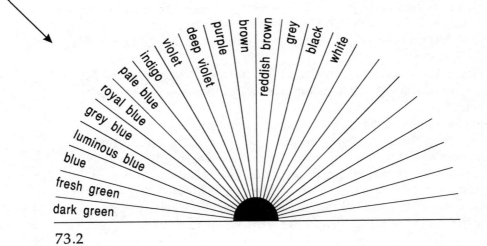

73.2

Figure 73. The colours of the aura

13. The chakras

Indian teachings first told us about the chakras. Chakras are physical, invisible energy-centres supplying us with cosmic energy. The word chakra comes from Sanskrit and means 'wheel'. Clairvoyants see chakras as funnel-shaped energy fields. They are also known as lotuses.

Through these chakras we obtain energy and life force from the cosmos. They can be dull and rather small, or large and full of vitality. This depends on the person's health and level of consciousness. There are seven major chakras, which relate to certain organs and parts of the body.

Furthermore, each chakra can be associated with a specific phase of development. This development proceeds from the root chakra upwards. The lower chakras turn slower and are related to the fundamental necessities of life. The higher chakras are associated with the ethereal, spiritual side of human experience.

With the help of the pendulum you can determine someone's level of spiritual development. The physical side is not looked at. Often you will discover that the higher chakras of a great many people are only partly developed. Our karmic and cosmic task is to concentrate ourselves on our higher consciousness and not on securing material possessions and the pleasure this may give. The seven chakras and their functions can be described as follows.

The root chakra is located at the base of the spine. It relates to primitive energy and basic survival needs. It is associated with being down-to-earth and looking for your own niche. It is the energy-centre for reproduction and the corresponding organs.

The sacral chakra is the centre for feelings of self-esteem, determination and self-preservation. This chakra governs the inner reproductive organs, the kidneys, bladder and circulation.

The solar plexus is responsible for the emotions and moods and 'digesting' impressions. It governs the stomach, liver, gall and spleen. The solar plexus is for willpower, preserving your own identity, and -if used wrongly - for power.

The heart chakra is the chakra of communication and creativity. It is also the link between what we experience and the outside world. This we express by our voice, which also belongs to this chakra. The corresponding organs and parts of the body are the throat, thyroid gland, and the arms. Breathing is also governed by this chakra.

The frontal chakra or 'third eye' is located between the eyes. Spiritual insight and realization are related to this chakra. It is also associated with psychic phenomena such as clairvoyance and telepathy. The development of this chakra indicates the level of spiritual awareness. The corresponding areas are the lower part of the brain, the eyes, ears, nose and a part of the nervous system.

The crown chakra is located at the top of the head. It is the chakra representing the highest insight, the link with the Absolute. It is also the chakra of cosmic unity, wisdom and enlightenment. It governs the upper part of the brain.

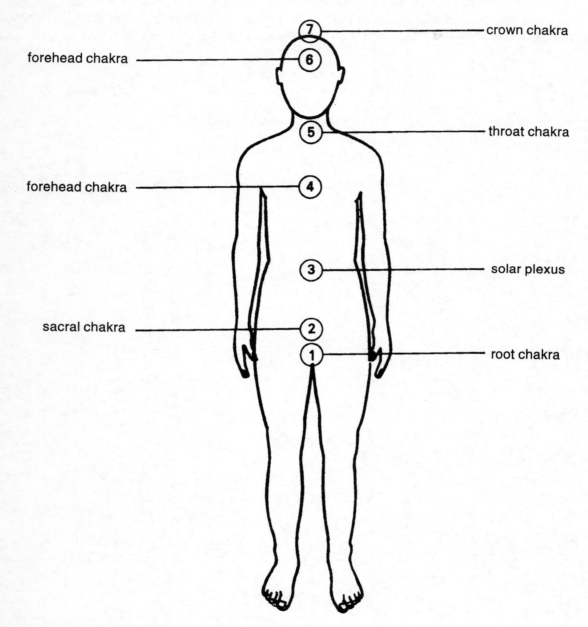

crown chakra

forehead chakra

throat chakra

forehead chakra

solar plexus

sacral chakra

root chakra

Figure 74. The chakras

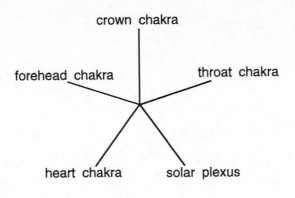

75.1

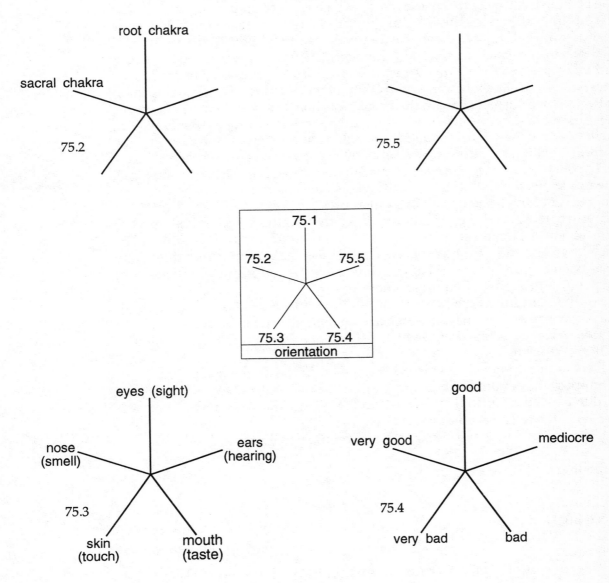

Figure 75. The vitality of the chakras

14. I Ching

The I Ching is an ancient Chinese oracle, which you can consult for advice. The sixty-four chapters of the I Ching describe what you are, and what your destination is. The advice may relate to any aspect of life. There are several methods to construct the sixty-four hexagrams: by using yarrow stalks, coins, and, of course, also by using the pendulum. Each hexagram is composed of six lines, which are either broken - - (yin) or unbroken — (yang). The changing lines, indicated by an asterisk, are in the process of changing into their counterparts; they represent the future. In the I Ching these lines are treated separately. You can determine the existing situation by looking at the signs as they are, and the coming situation by turning the changing lines into their counterparts.

To construct a hexagram , you start with the bottom line. This is the first line. The top line is the sixth (fig.76). You can also use the pendulum and the large diagrams to obtain the hexagrams more directly. To find out which lines are changing lines, hold the pendulum above the numbers 1 to 6. The number indicated by the pendulum will be the number of the changing line. So 1 will be the bottom line and 6 the top line. We can also ask the pendulum whether there is a second or possibly third changing line. If the pendulum indicates 'yes', use the pendulum again to find these other lines.

For the meaning of the hexagrams and the explanatory text, please consult one of the many books that are published on this subject. For instance, *I Ching, a new interpretation for modern times* by Sam Reifler or *I Ching, the book of changes* by Richard Wilhelm.

Here is an example of how to consult the I Ching with the pendulum.

Suppose the pendulum indicates the first line to be —*, the second line - -, the third line - -*, the fourth line —, the fifth line —,and the sixth line - -. You will then have the following diagram:

	existing situation	coming situation
6th line	- -	- -
5th line	— *	—
4th line	—	—
3rd line	- -*	—
2nd line	- -	- -
1st line	— *	- -

no.17 Following / Sui   no.31 Influence (Wooing) / Hsien

Always consult the I Ching and read carefully how to use the indications for yourself and others.

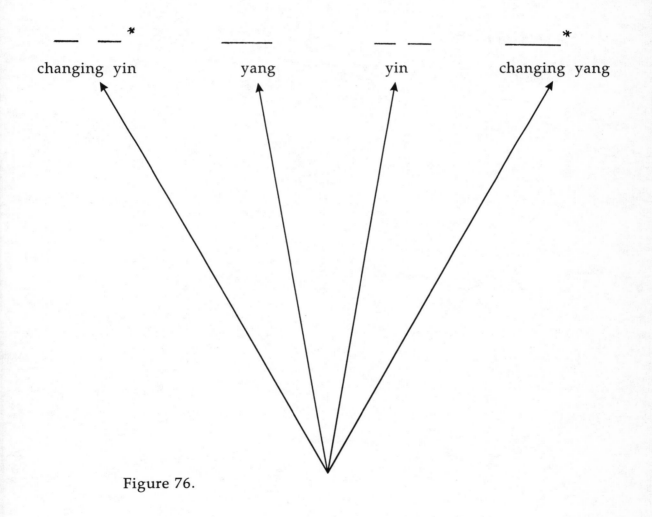

Figure 76.

upper trigrams

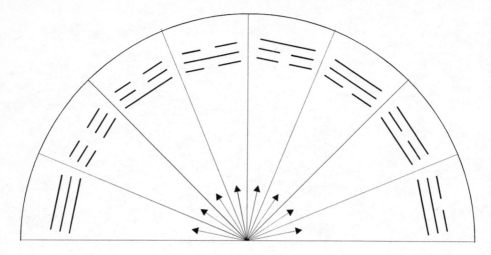

77.1

lower trigrams

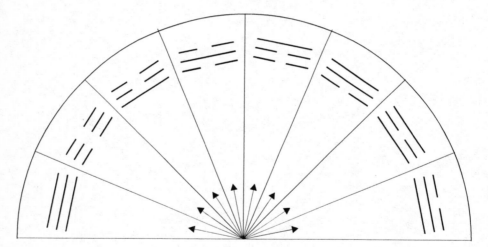

77.2

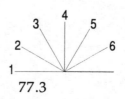

77.3

Figure 77. The trigrams

114

trigrams

upper

lower							
1	11	34	5	26	9	14	43
12	2	16	8	23	20	35	45
25	24	51	3	27	42	21	17
6	7	40	29	4	59	64	47
33	15	62	39	52	53	56	31
44	46	32	48	18	57	50	28
13	36	55	63	22	37	30	49
10	19	54	60	41	61	38	58

Figure 78. The Hexagram

15. Weather

People always wonder what the weather will be like. It may be nice today, but what about tomorrow? Can I give this garden party or will it rain? Will it be dry next week? What about that cycling tour, visit to the amusement park, or walk on the beach? What about the holidays? To determine whether it will rain or not on a certain day, focus your mind on that day or make a note. Hold the pendulum above the centremost figure and ask: What kind of weather will it be on ... (date)? The pendulum will indicate whether you will need an umbrella, warm clothing or sun tan oil. If you have not yet fixed a date for a certain festivity, but you would like it to be on a nice day, then use the diagrams of the chapters on 'Astrology'. There you will find entries for day and month, which will help you to determine the right day. Sailors, surfers and other people who rely on wind can use the pendulum to determine the wind-direction and wind force.

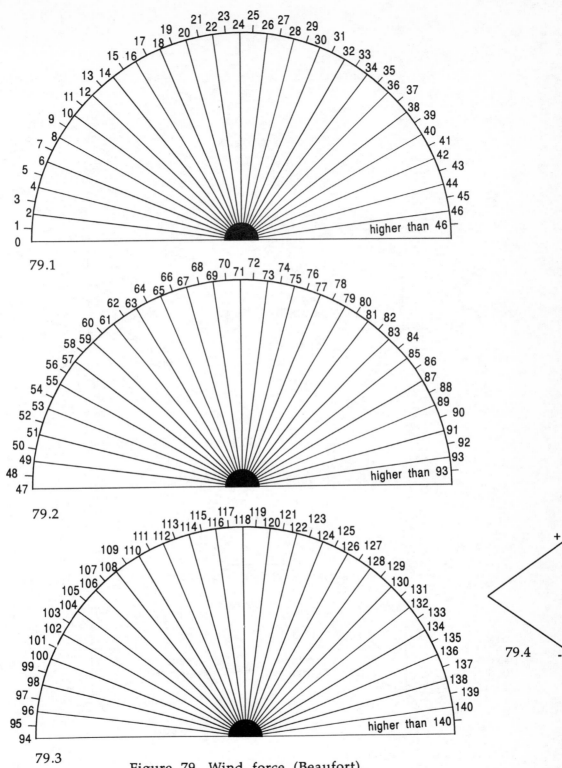

79.1

79.2

79.3

79.4

Figure 79. Wind force (Beaufort),
precipitation (mm), temperature (Celsius)

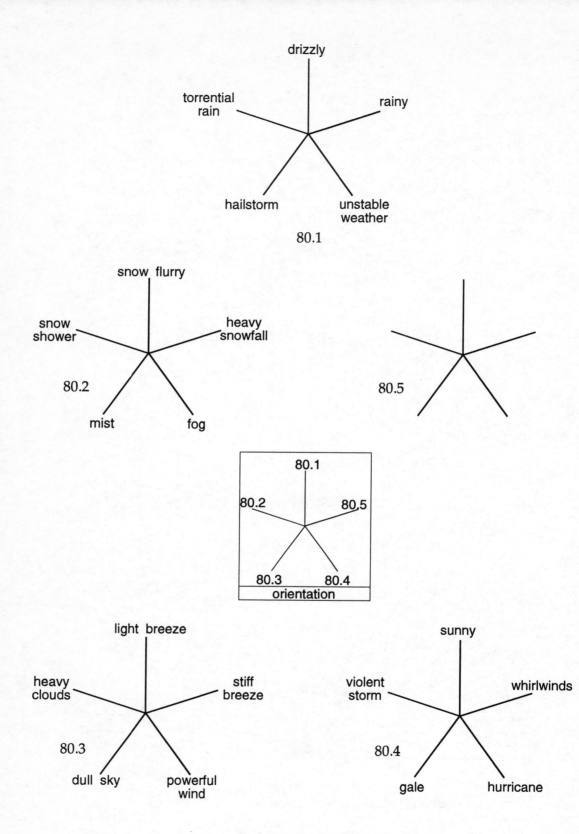

Figure 80. Weather conditions

16. Holidays

Planning your holidays can be great fun. Where to go, with whom, and how to get there? People who would like to go somewhere else for a change will find the following diagrams helpful. These provide a wide range of possibilities for very enjoyable holidays.

If you are not travelling alone, you may also let your companions use the pendulum to find your favourite destination. If you can plan your holidays whenever you like, you might want to use the pendulum to determine a suitable period in the off-season. The diagrams are also useful for fixing times of departures and dates.

The previous chapter on 'Weather' may also prove helpful in planning your holidays. If you are more attracted to an active holiday or a holiday course, have a look at the chapter on 'Sports and leisure activities'. The diagrams belonging to this chapter can also be used to trace someone who has gone missing. See also chapter 22 'Lost objects and missing persons'.

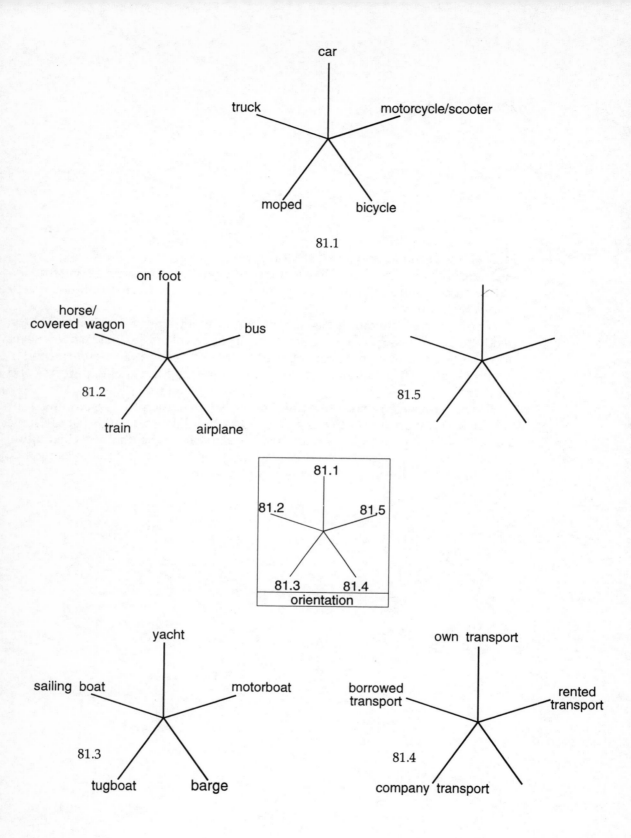

Figure 81. Transport

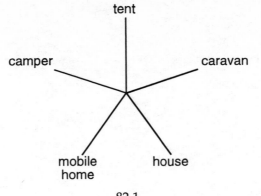

82.1

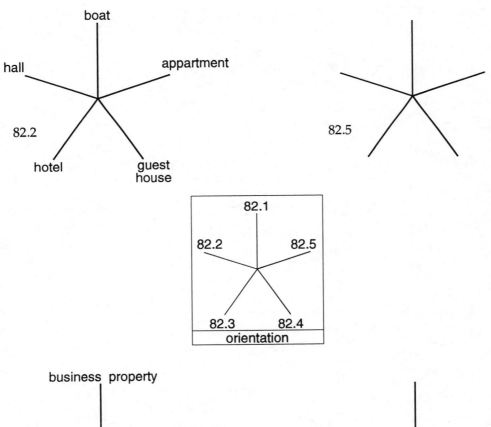

82.2

82.5

82.3

82.4

orientation

Figure 82. Accomodation

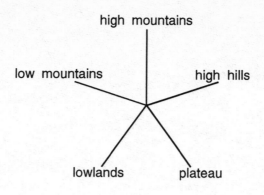

high mountains

low mountains   high hills

lowlands   plateau

83.1

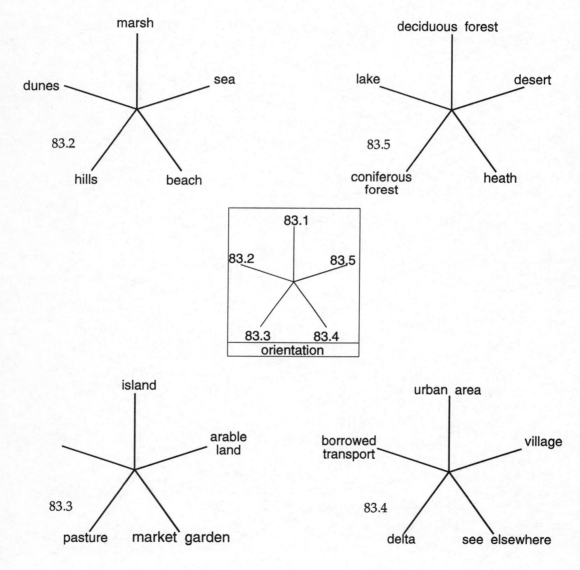

marsh

dunes   sea

83.2

hills   beach

deciduous forest

lake   desert

83.5

coniferous
forest   heath

83.1

83.2   83,5

83.3   83.4
orientation

island

arable
land

83.3

pasture   market garden

urban area

borrowed
transport   village

83.4

delta   see elsewhere

Figure 83. Landscapes

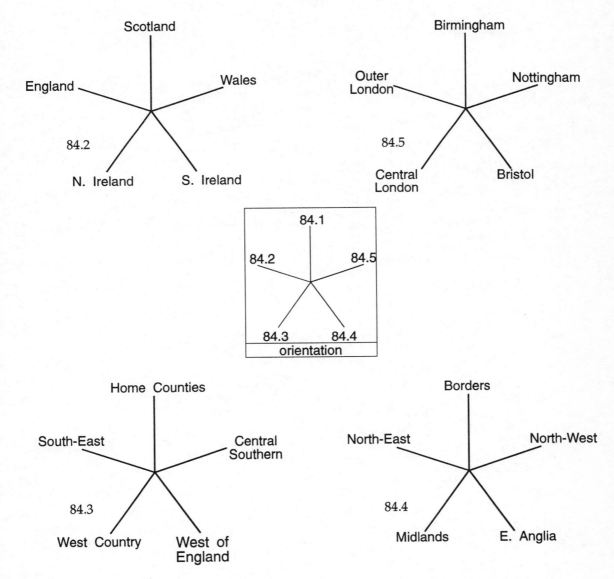

Figure 84. Places in the U.K. and Ireland

123

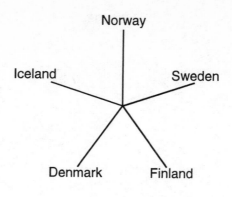

85.1

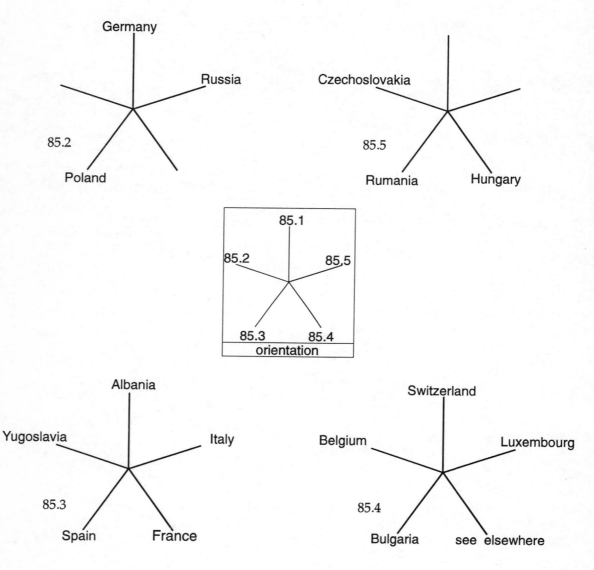

85.2

85.5

85.3

85.4

orientation

Figure 85. European countries

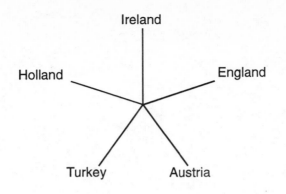

Ireland

Holland England

Turkey Austria

85.6

Scotland

Portugal Greece

85.7

Liechtenstein Andorra

Iraq

Syria Iran

85.8

Jordan Saudi Arabia

85.6

85.7 85.8

86.1 86.2

orientation

Europe

Asia Australasia

86.1

America Africa

Israel

Yemen Egypt

86.2

Kuwait Middle East

Figure 86. Continents and Countries

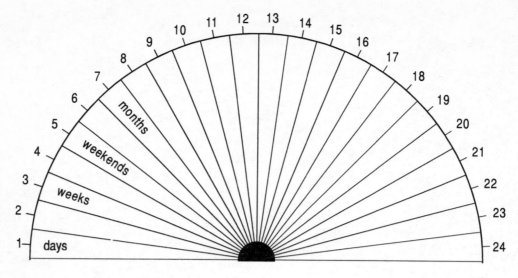

87.1

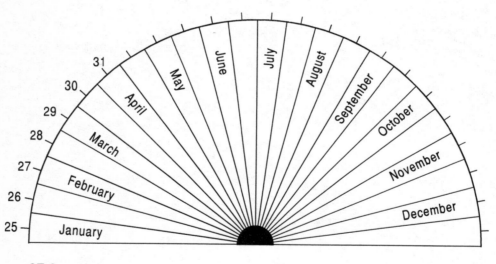

87.2

Figure 87. Length of holidays and date of departure

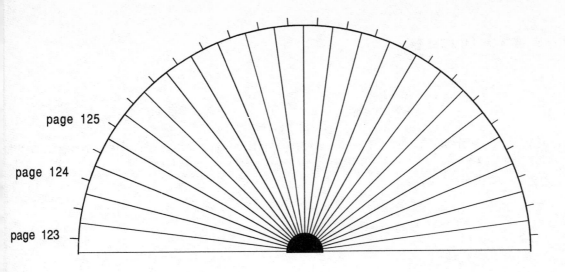

page 125

page 124

page 123

Figure 88. Number of destination page listing

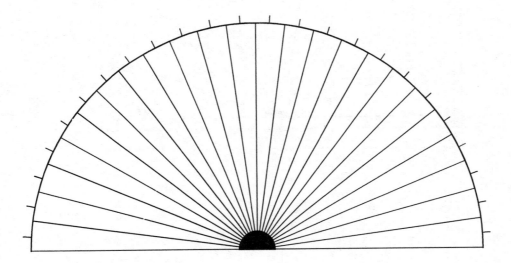

17. Sports and leisure

Nowadays, people have more time for recreation. Fortunately enough, there are many possibilities to enjoy oneself. Here are some examples to choose from.

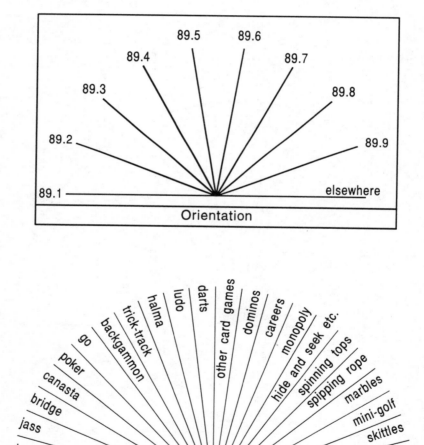

89.1 Games

Figure 89. Leisure activities

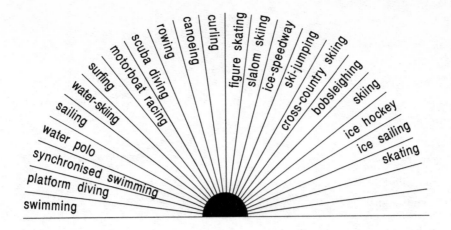

89.2 Water and wintersports

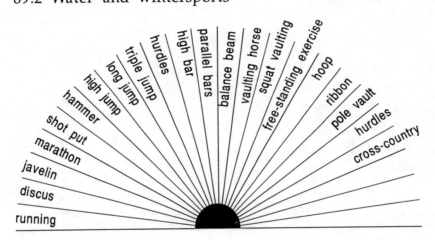

89.3 Athletics

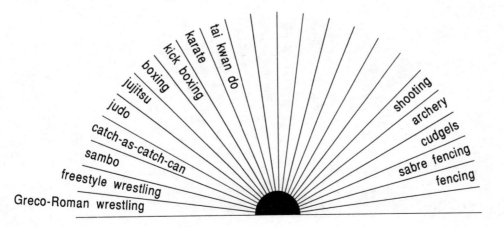

89.4 Martial arts

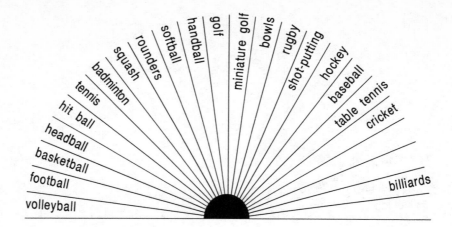

89.5 Ball games

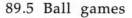

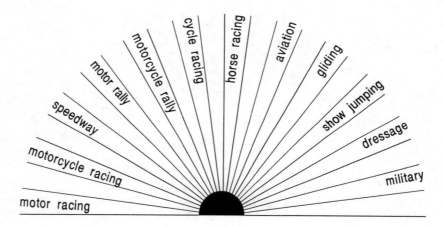

89.6 Other sports

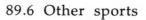

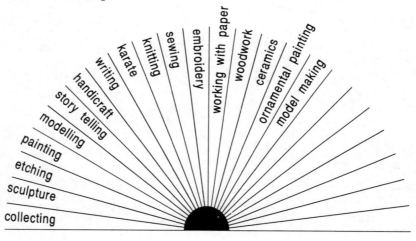

89.7 Creative hobbies

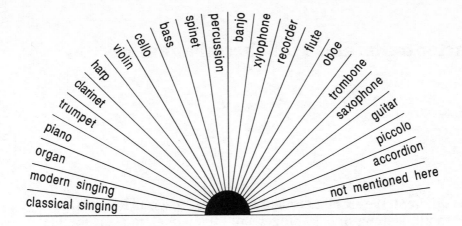

89.8 Music

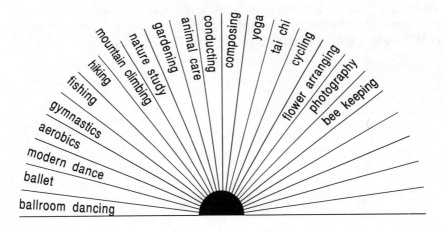

89.9 Other leisure activities

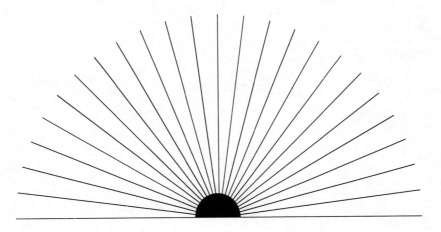

18. Character traits, areas for special attention

Each person is different. We are all unique. This is because we all have different characters. Without wishing to pass judgement on anyone, you can say that certain character traits are positive, negative or even very important for certain tasks or occupations. Certain character traits can either help or hinder us. They affect the way we act and the relationship we have with ourselves and others.

Self-knowledge provides us with an opportunity to change, to improve ourselves. Figure 92 'areas for special attention' may be helpful in this respect.

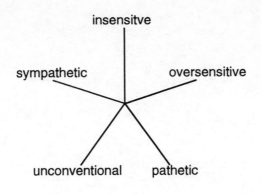

90.1

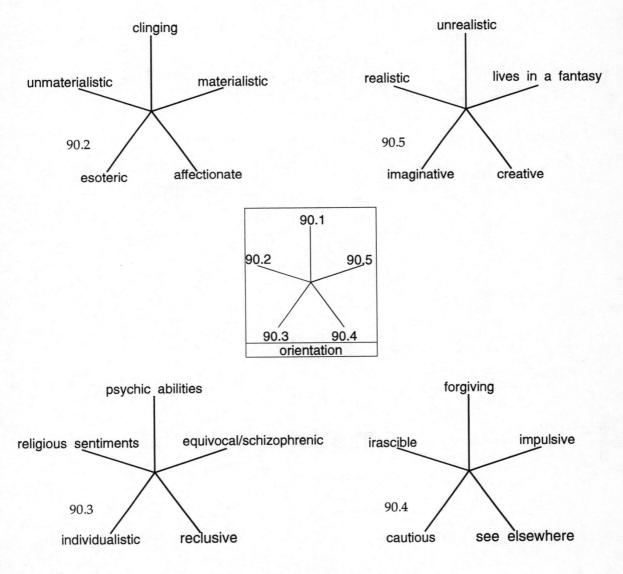

Figure 90. Character and type

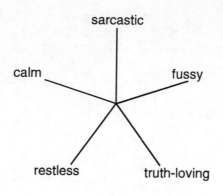

sarcastic

calm   fussy

restless   truth-loving

90.6

spontaneous

fair   tolerant

90.7

unfair   oversentsitive to injustice

strong-willed

born loser   badloser

90.10

easily discouraged   timid

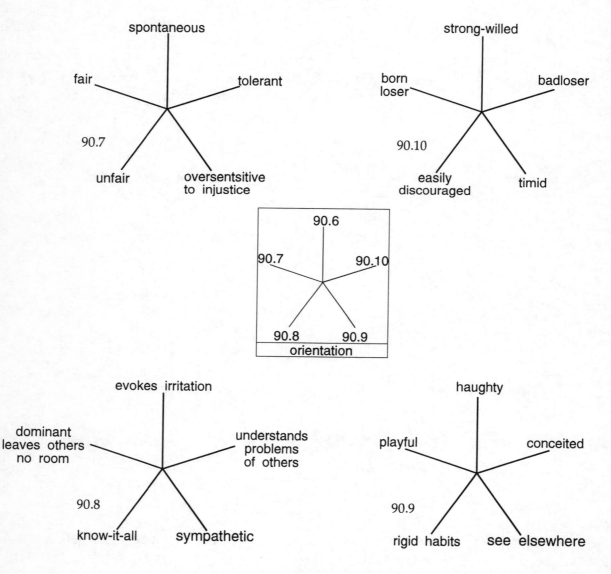

90.6

90.7   90.10

90.8   90.9

orientation

evokes irritation

dominant leaves others no room   understands problems of others

90.8

know-it-all   sympathetic

haughty

playful   conceited

90.9

rigid habits   see elsewhere

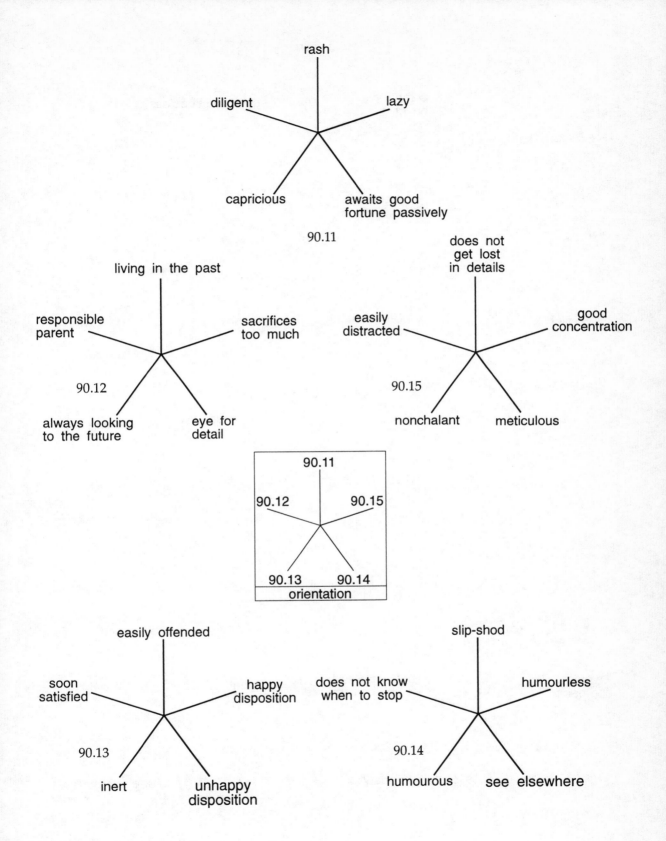

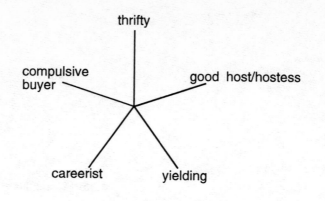

thrifty

compulsive buyer — good host/hostess

careerist — yielding

90.16

generous

careful — likes action/excitement

90.17

careless — ruthless

neurotic

psychotic — cruel

90.20

peacemaker — reckless

90.16

90.17 — 90.20

90.18 — 90.19

orientation

good approach

wrong approach — acts discreetly

90.18

polite — acts conspiculously

intellectual approach

brute — emotional approach

90.19

likes good manners — see elsewhere

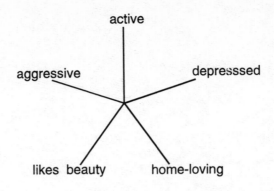

active

aggressive   depresssed

likes beauty   home-loving

90.21

cautious

loves plain   loves warm
colours   colours

90.22

loves pastel   loves bright
shades   colours

impressionable

epicurean   easily
                   discouraged

90.25

merciless   thoughtless

90.21

90.22   90.25

90.23   90.24

orientation

likes sports

caring   prodigal

90.23

miser   squanderer

likes reading

likes silence   does not like
                       silence

90.24

nature lover   see elsewhere

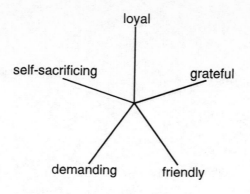

loyal

self-sacrificing grateful

demanding friendly

90.26

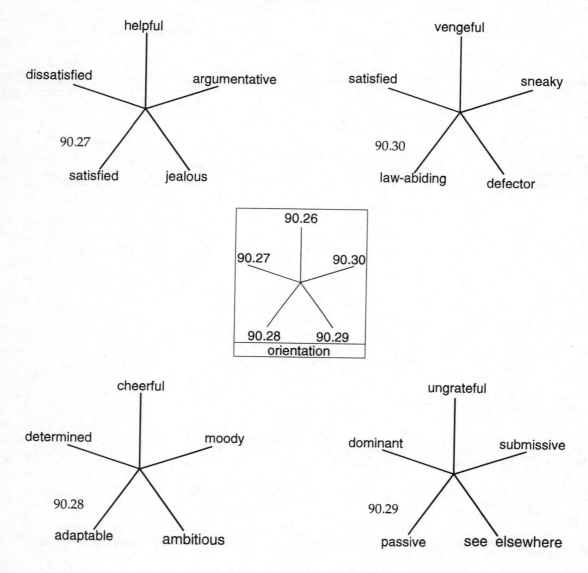

helpful

dissatisfied argumentative

90.27

satisfied jealous

vengeful

satisfied sneaky

90.30

law-abiding defector

90.26

90.27 90.30

90.28 90.29
orientation

cheerful

determined moody

90.28

adaptable ambitious

ungrateful

dominant submissive

90.29

passive see elsewhere

138

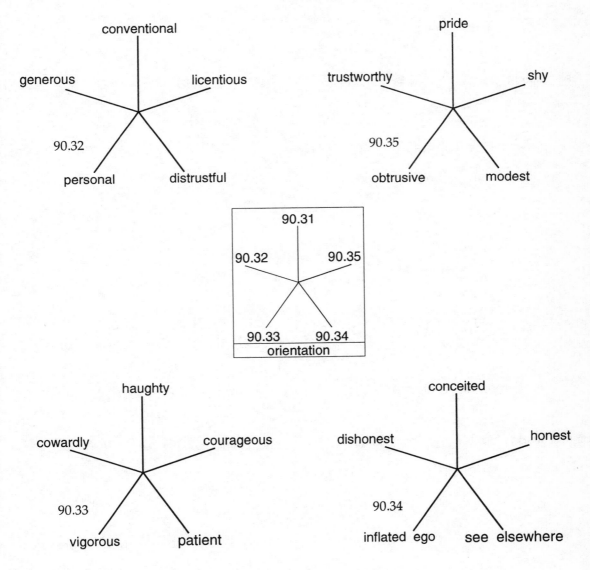

vain

self-satisfied   egotistic

altruistic   anxious

90.31

conventional

generous   licentious

90.32

personal   distrustful

pride

trustworthy   shy

90.35

obtrusive   modest

90.31

90.32   90.35

90.33   90.34
orientation

haughty

cowardly   courageous

90.33

vigorous   patient

conceited

dishonest   honest

90.34

inflated ego   see elsewhere

139

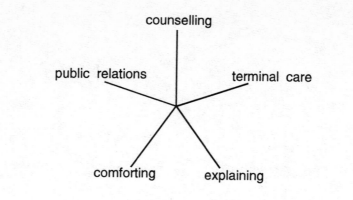

counselling

public relations   terminal care

comforting   explaining

91.1

encouraging

stimulating   management

91.2

consoling   negotiating

91.5

91.1

91.2   91.5

91.3   91.4

orientation

insight into character

spiritual leadership   organisation

91.3

partnership

mental care   happiness

91.4

theories   pragmatism

Figure 91. Talents

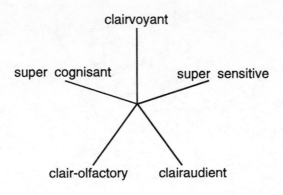

clairvoyant

super cognisant   super sensitive

clair-olfactory   clairaudient

91.6

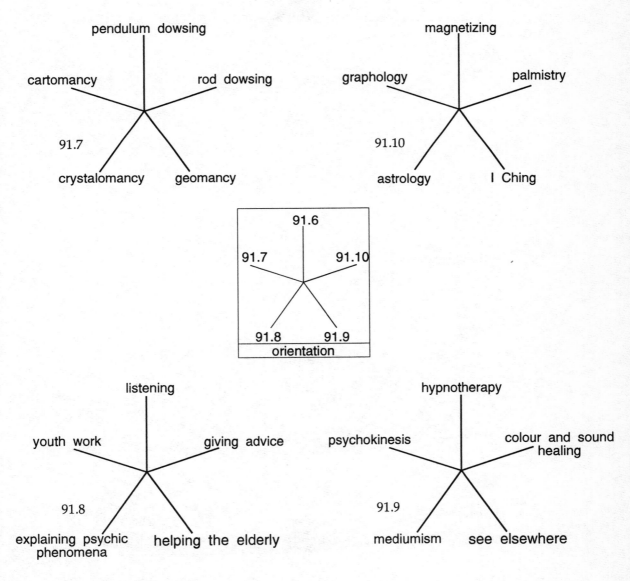

pendulum dowsing

cartomancy   rod dowsing

91.7

crystalomancy   geomancy

magnetizing

graphology   palmistry

91.10

astrology   I Ching

91.6

91.7   91.10

91.8   91.9

orientation

listening

youth work   giving advice

91.8

explaining psychic   helping the elderly
phenomena

hypnotherapy

psychokinesis   colour and sound
healing

91.9

mediumism   see elsewhere

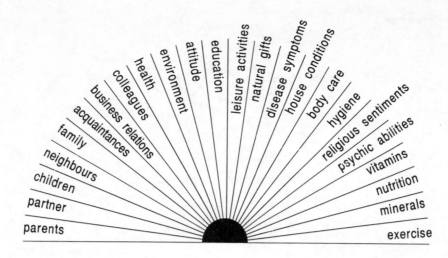

92.1

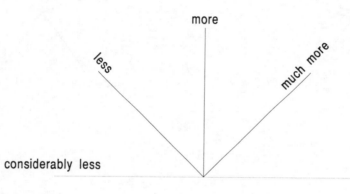

92.2 More/less attention

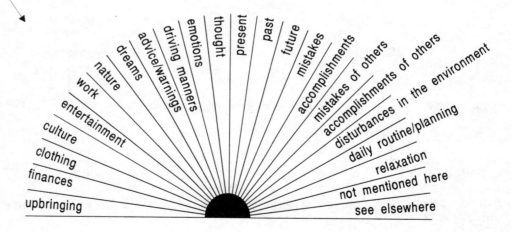

92.3

Figure 92. Areas for special attention

19. Headlines

Sometimes the news will jolt us out of our complacency. Thanks to the radio, television and newspaper, the world is closer to home. The pendulum can also be used to find out how long a certain situation will last or how it will develop. With the help of figure 93 you can use the pendulum to determine important events. Figure 94 represents the areas of activity.

Figure 95 may help to determine how much time will pass between 'now' and the moment something will happen and how long this new situation will last.

First use the pendulum to find the day, week, month, year, or century in figure 95.1, and then the corresponding number (fig.95.1-95.5).

For countries, use the diagrams in chapter 16.

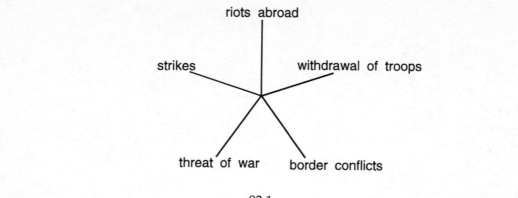

93.1

change of power
nuclear power issues
natural disasters

93.2

environmental pollution
environmental disasters

breakthrough
disaster
special event

93.5

failure
success

93.1
93.2   93.5
93.3   93.4
orientation

recognition of important figure
kidnapping/ hostage-taking
new discovery

93.3

conflicts
death of an important figure

decrease in imports
decrease in exports
see elsewhere

93.4

increase in exports
increase in imports

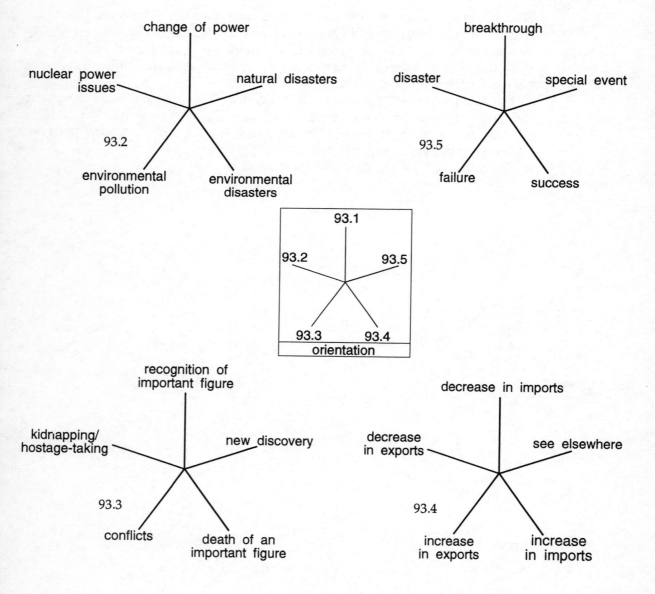

144

Figure 93. Important events

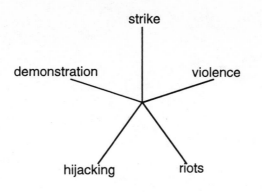

strike

demonstration

violence

hijacking

riots

93.6

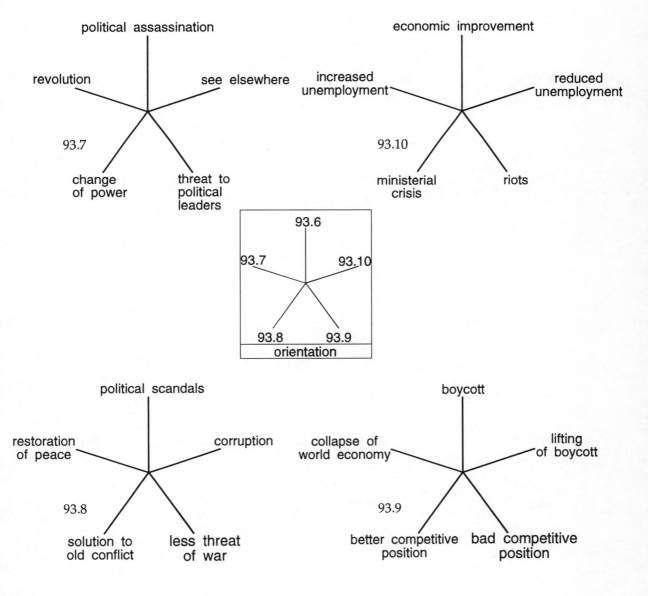

political assassination

revolution

see elsewhere

93.7

change
of power

threat to
political
leaders

economic improvement

increased
unemployment

reduced
unemployment

93.10

ministerial
crisis

riots

93.6

93.7

93.10

93.8

93.9

orientation

political scandals

restoration
of peace

corruption

93.8

solution to
old conflict

less threat
of war

boycott

collapse of
world economy

lifting
of boycott

93.9

better competitive
position

bad competitive
position

birth

marriage divorce

engagement new partner

93.11

dismissal

promotion change of job

93.12

progress no recognition

fire

threats vandalism

93.15

forgery destruction

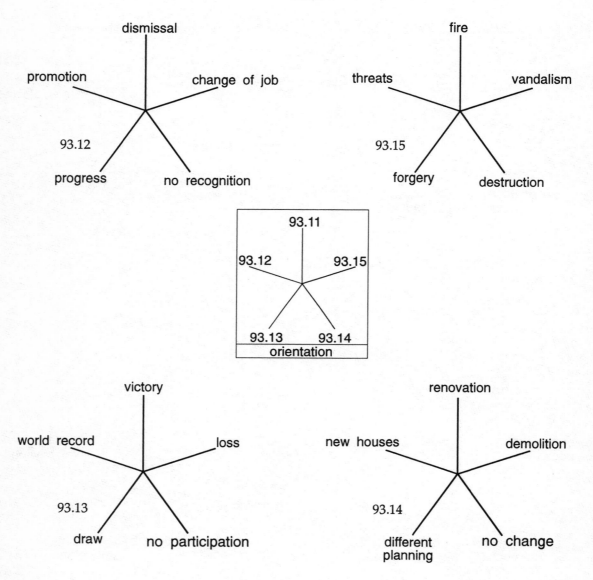

93.11

93.12 93.15

93.13 93.14

orientation

victory

world record loss

93.13

draw no participation

renovation

new houses demolition

93.14

different planning no change

146

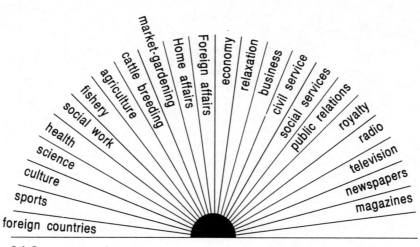

94.2

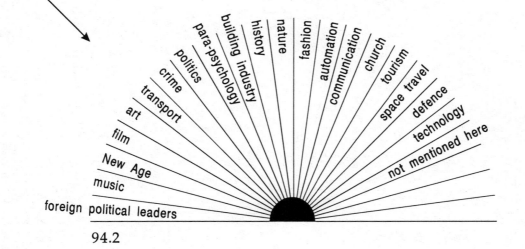

94.2

Figure 94. Aspects of life

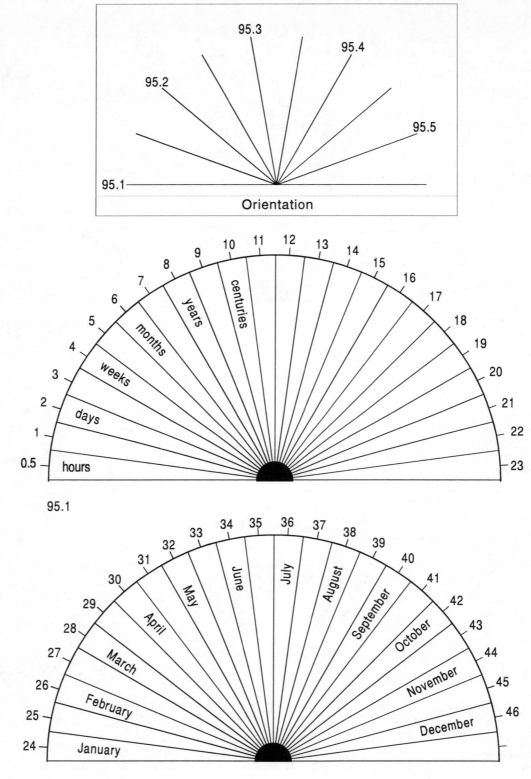

Figure 95. Point and length of time

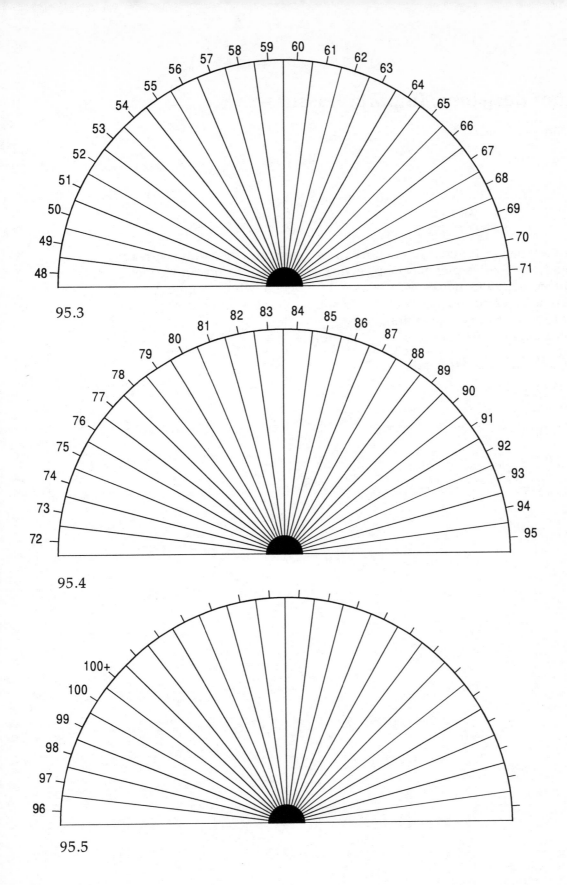

95.3

95.4

95.5

149

20. Minor decisions, Important issues

Life is constant change. We change, circumstances change and thereupon we react with changes, etc. This chapter may help you to take minor decisions and to make adjustments in daily life. It is especially useful for people who have the feeling that something is wrong, but cannot trace its origin. It is also suitable for people who want to give life a new direction.

Please formulate your questions very carefully.

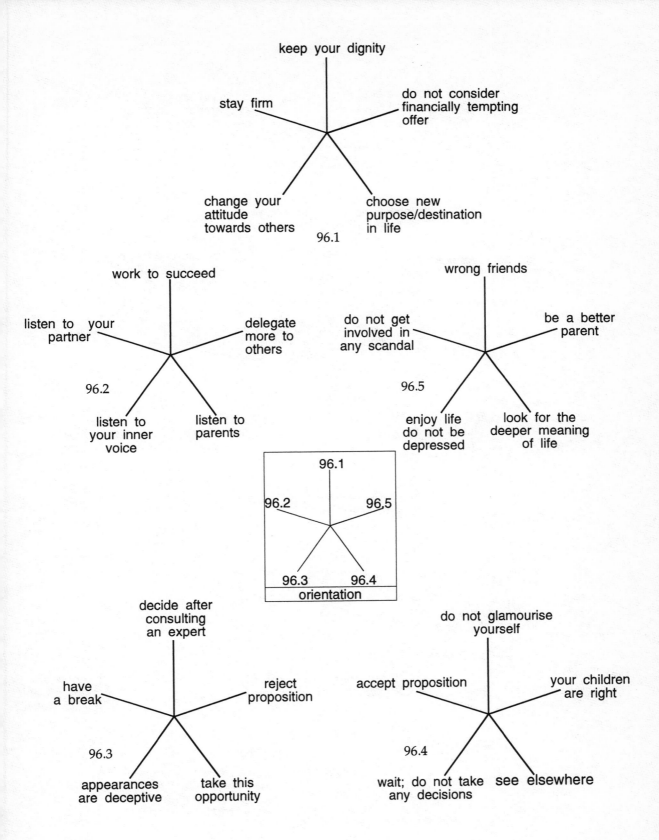

keep your dignity

stay firm

do not consider
financially tempting
offer

change your
attitude
towards others

choose new
purpose/destination
in life

96.1

work to succeed

listen to your
partner

delegate
more to
others

96.2

listen to
your inner
voice

listen to
parents

wrong friends

do not get
involved in
any scandal

be a better
parent

96.5

enjoy life
do not be
depressed

look for the
deeper meaning
of life

96.1

96.2

96,5

96.3

96.4

orientation

decide after
consulting
an expert

have
a break

reject
proposition

96.3

appearances
are deceptive

take this
opportunity

do not glamourise
yourself

accept proposition

your children
are right

96.4

wait; do not take
any decisions

see elsewhere

151

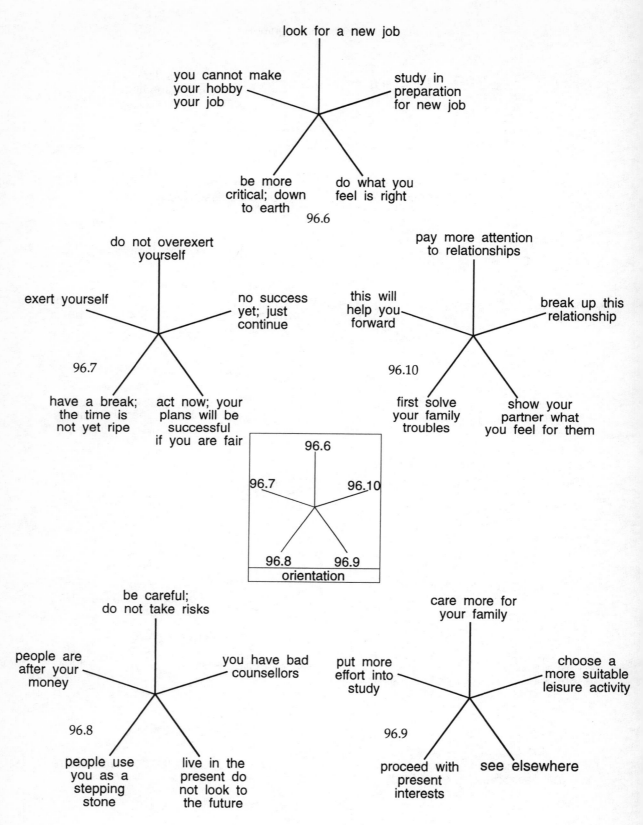

look for a new job

you cannot make your hobby your job

study in preparation for new job

be more critical; down to earth

do what you feel is right

96.6

do not overexert yourself

exert yourself

no success yet; just continue

96.7

have a break; the time is not yet ripe

act now; your plans will be successful if you are fair

pay more attention to relationships

this will help you forward

break up this relationship

96.10

first solve your family troubles

show your partner what you feel for them

96.6

96.7   96.10

96.8   96.9

orientation

be careful; do not take risks

people are after your money

you have bad counsellors

96.8

people use you as a stepping stone

live in the present do not look to the future

care more for your family

put more effort into study

choose a more suitable leisure activity

96.9

proceed with present interests

see elsewhere

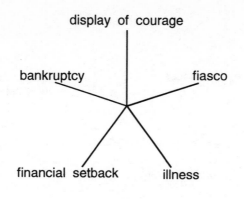

97.1

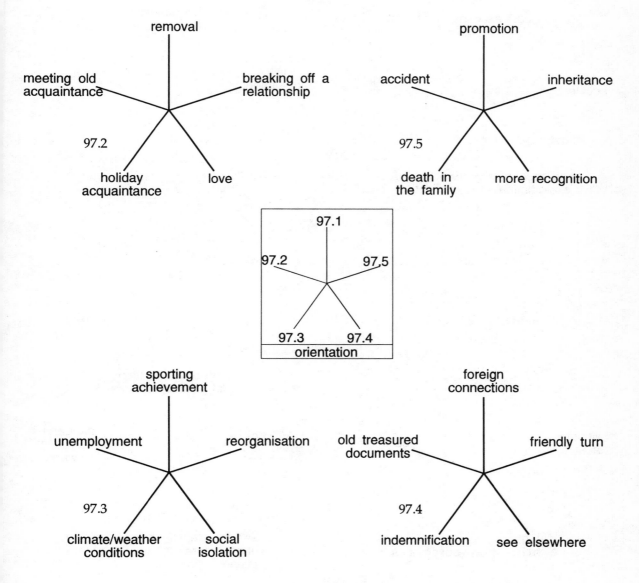

97.2

97.3

97.5

97.4

orientation

Figure 97. Crucial events

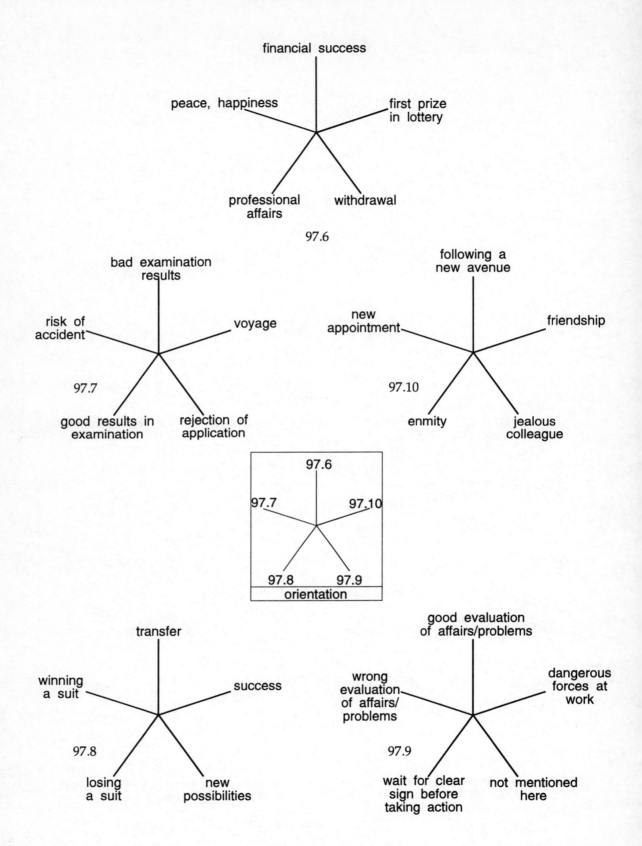

financial success

peace, happiness

first prize
in lottery

professional
affairs

withdrawal

97.6

bad examination
results

risk of
accident

voyage

97.7

good results in
examination

rejection of
application

following a
new avenue

new
appointment

friendship

97.10

enmity

jealous
colleague

97.6

97.7

97.10

97.8

97.9

orientation

transfer

winning
a suit

success

97.8

losing
a suit

new
possibilities

good evaluation
of affairs/problems

wrong
evaluation
of affairs/
problems

dangerous
forces at
work

97.9

wait for clear
sign before
taking action

not mentioned
here

21. Professions and business

This chapter is subdivided into three categories: professions, character traits in relation to business, and business life in general.

Figure 98 (two pages) lists a great many professions. This figure can be used when you wish to choose a career and to see which profession might suit you.

Figure 99 lists a great many character traits. These are either positive or negative with regard to certain business professions. Character traits can be important when you are applying for a job, hoping for promotion, or being transferred, etc.

It might also be interesting to see whether you are doing well in your present job.

Figure 100 gives some business advice. Decisions can cause small or great changes, and those may affect a company's position positively or negatively.

Although it is not recommended to trust the pendulum blindly in everything, it can sometimes make you look at things from a different angle. Always check if the indications of the pendulum are actually feasible or whether your unconscious wishes and desires have perhaps influenced the results.

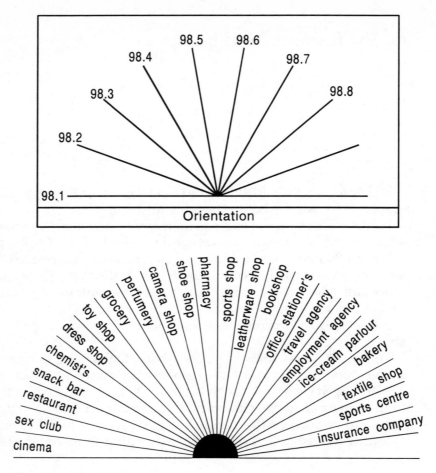

Figure 98.1 Owner, manager, employee

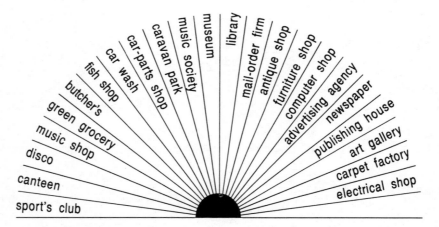

98.2 Owner, manager, employee

Figure 98. Professions

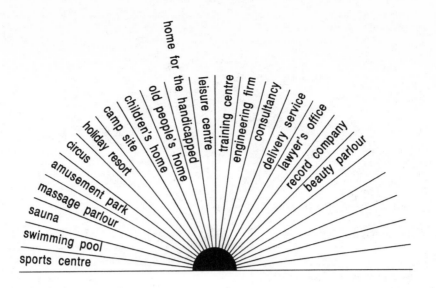

98.3 Owner, manager employee

home for the handicapped, leisure centre, training centre, engineering firm, consultancy, delivery service, lawyer's office, record company, beauty parlour, old people's home, children's home, camp site, holiday resort, circus, amusement park, massage parlour, sauna, swimming pool, sports centre

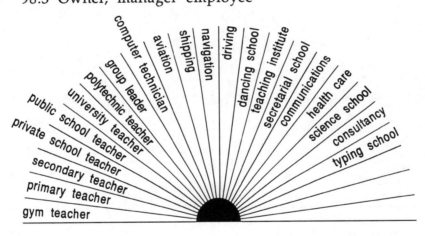

98.4 Education

computer technician, aviation, shipping, navigation, driving, dancing school, teaching institute, secretarial school, communications, health care, science school, consultancy, typing school, group leader, polytechnic teacher, university teacher, public school teacher, private school teacher, secondary teacher, primary teacher, gym teacher

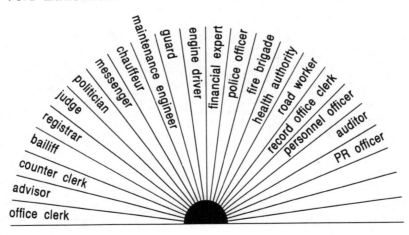

98.5 Public service

maintenance engineer, guard, engine driver, financial expert, police officer, fire brigade, health authority, road worker, record office clerk, personnel officer, auditor, PR officer, chauffeur, messenger, politician, judge, registrar, bailiff, counter clerk, advisor, office clerk

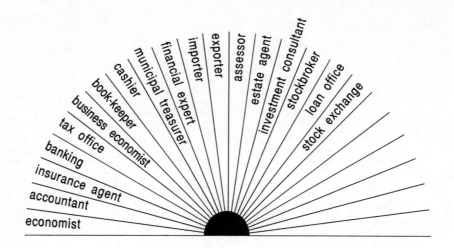

98.6 Financial sector

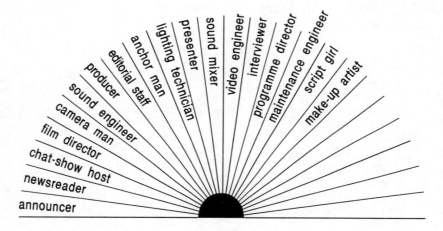

98.7 Radio and television

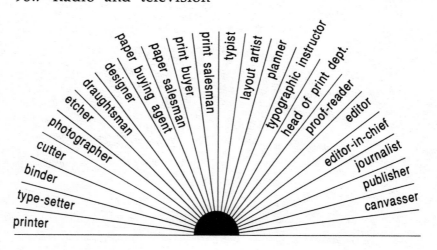

98.8 Printing and allied trades

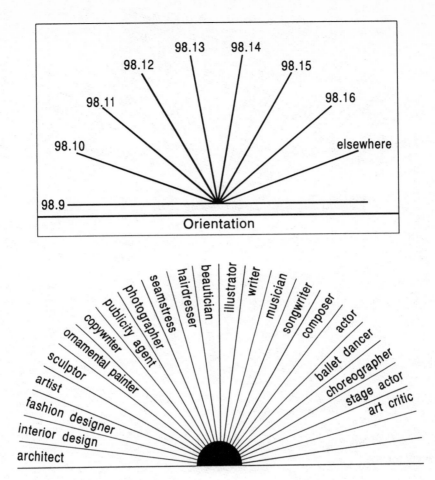

98.9 Culture and the arts

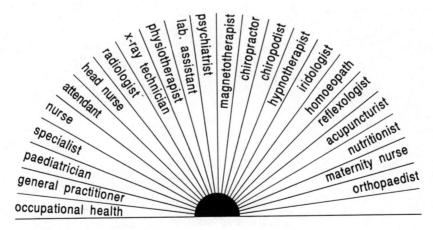

98.10 Health care

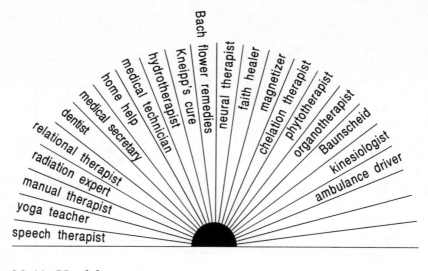

98.11 Health care

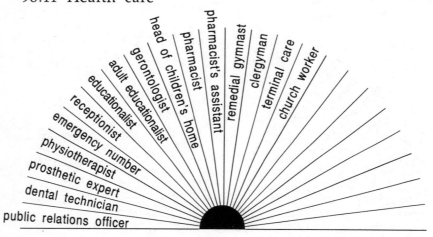

98.12 Health care

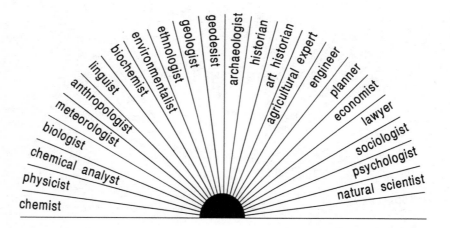

98.13 Science

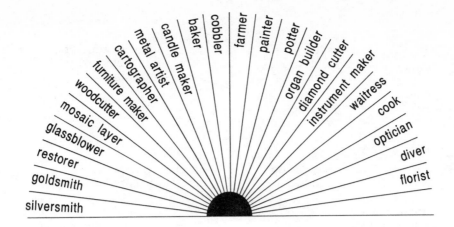

98.14 Trades

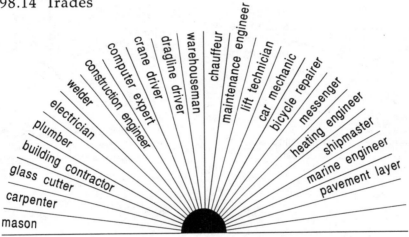

98.15 Other trades and professions

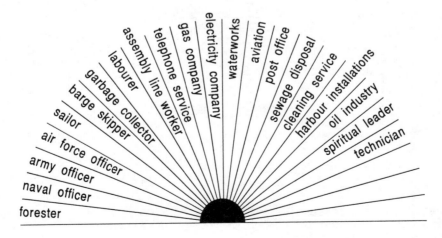

98.16 Other trades and professions

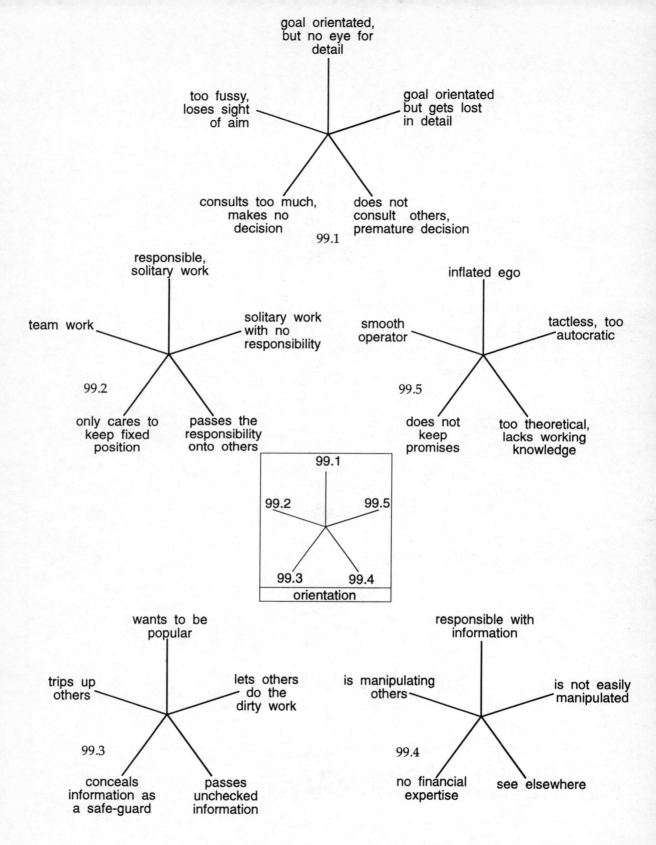

Figure 99. Important character traits for business and professions

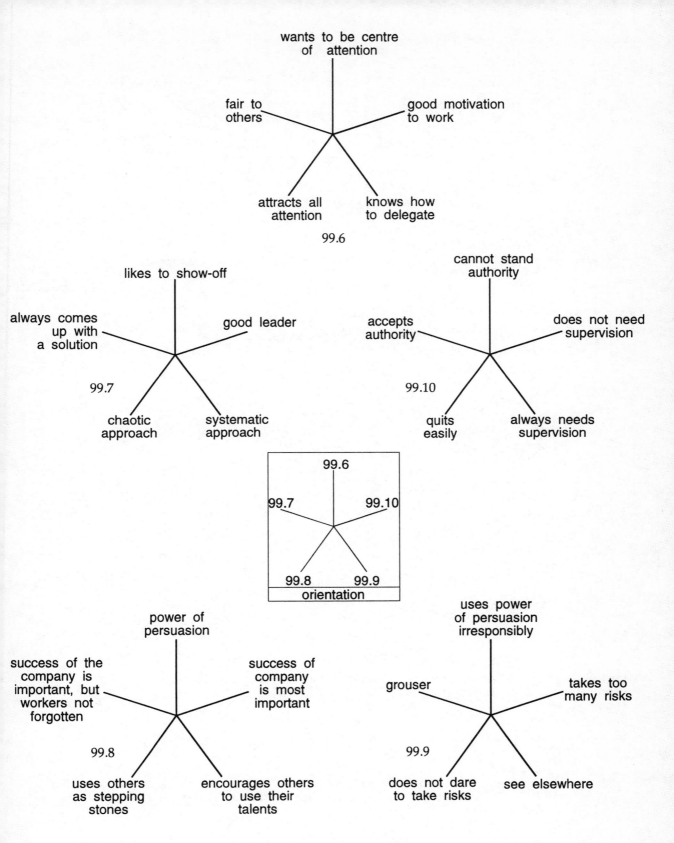

wants to be centre
of attention

fair to others — good motivation to work

attracts all attention — knows how to delegate

99.6

likes to show-off

always comes up with a solution — good leader

99.7

chaotic approach — systematic approach

cannot stand authority

accepts authority — does not need supervision

99.10

quits easily — always needs supervision

99.6

99.7   99.10

99.8   99.9

orientation

power of persuasion

success of the company is important, but workers not forgotten — success of company is most important

99.8

uses others as stepping stones — encourages others to use their talents

uses power of persuasion irresponsibly

grouser — takes too many risks

99.9

does not dare to take risks — see elsewhere

163

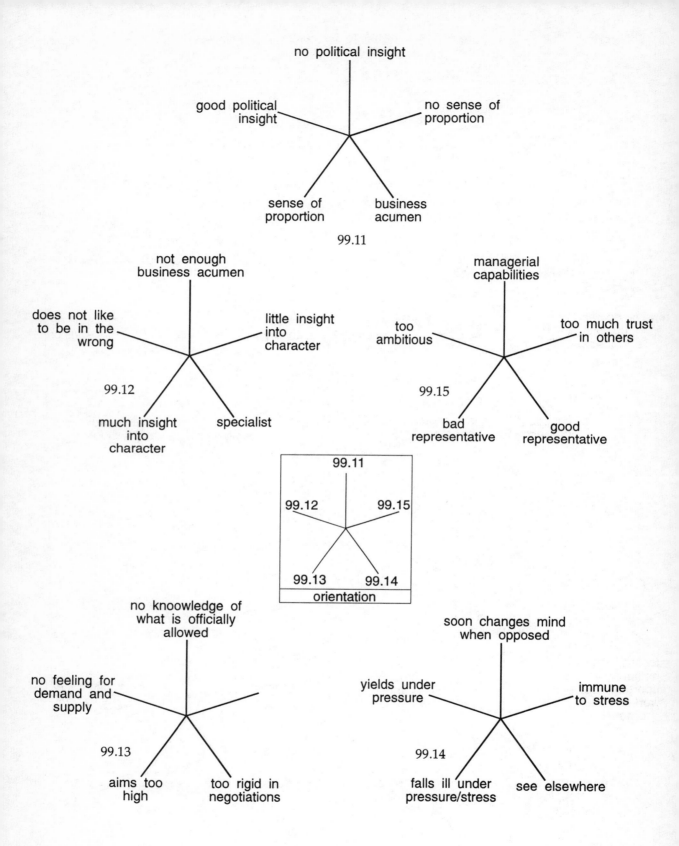

no political insight

good political insight

no sense of proportion

sense of proportion

business acumen

99.11

not enough business acumen

does not like to be in the wrong

little insight into character

99.12

much insight into character

specialist

managerial capabilities

too ambitious

too much trust in others

99.15

bad representative

good representative

99.11

99.12   99.15

99.13   99.14

orientation

no knoowledge of what is officially allowed

no feeling for demand and supply

99.13

aims too high

too rigid in negotiations

soon changes mind when opposed

yields under pressure

immune to stress

99.14

falls ill under pressure/stress

see elsewhere

164

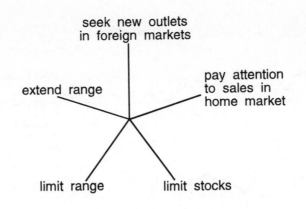

seek new outlets
in foreign markets

extend range

pay attention
to sales in
home market

limit range

limit stocks

100.1

more investments

less
investments

more staff

100.2

less automize

sell the
company

expand stocks

centralise

decentralise

100.5

more
responsibility
for staff

more
supervision
over staff

100.1

100.2

100.5

100.3

100,4

orientation

increase speed

use other
methods of
fabrication

automate

100.3

do not
automate

stop automation

stop mechanization

more
mechanization

choose new
product

100.4

reduce liabilities

see elsewhere

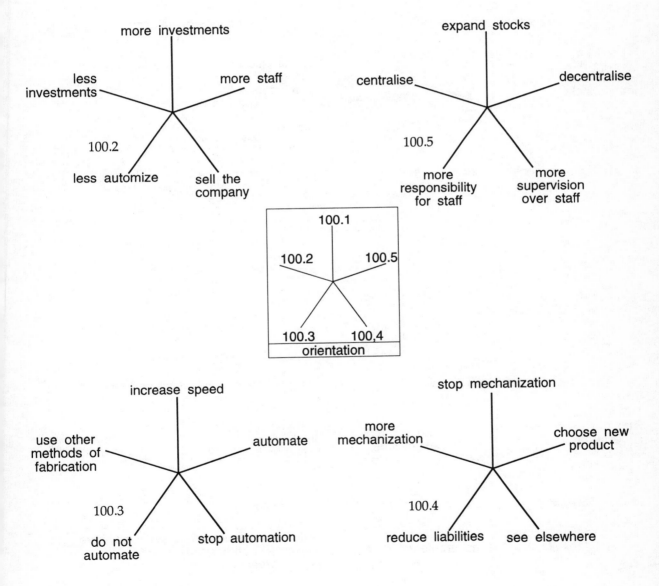

Figure 100. Business advice

165

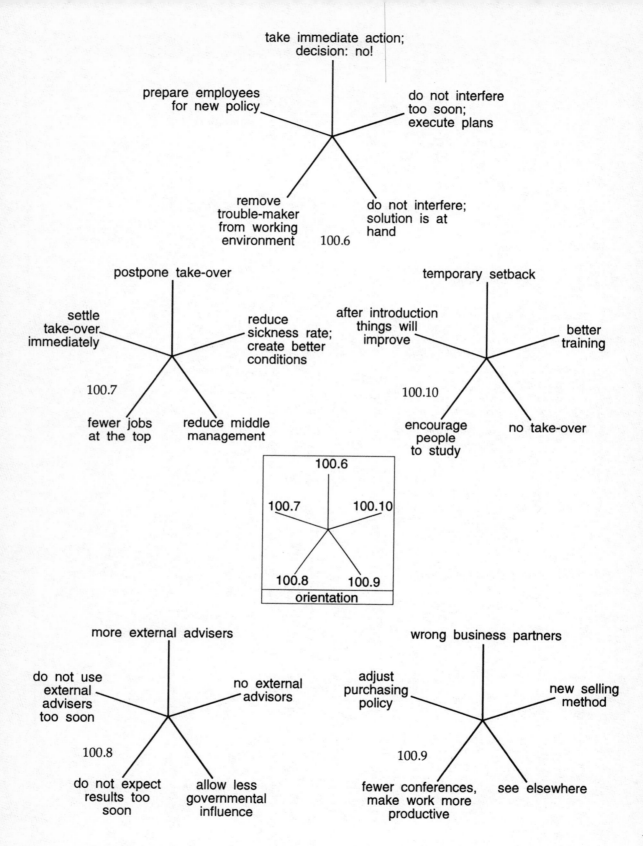

take immediate action;
decision: no!

prepare employees
for new policy

do not interfere
too soon;
execute plans

remove
trouble-maker
from working
environment   100.6

do not interfere;
solution is at
hand

postpone take-over

settle
take-over
immediately

reduce
sickness rate;
create better
conditions

100.7

fewer jobs
at the top

reduce middle
management

temporary setback

after introduction
things will
improve

better
training

100.10

encourage
people
to study

no take-over

100.6

100.7   100.10

100.8   100.9

orientation

more external advisers

do not use
external
advisers
too soon

no external
advisors

100.8

do not expect
results too
soon

allow less
governmental
influence

wrong business partners

adjust
purchasing
policy

new selling
method

100.9

fewer conferences,
make work more
productive

see elsewhere

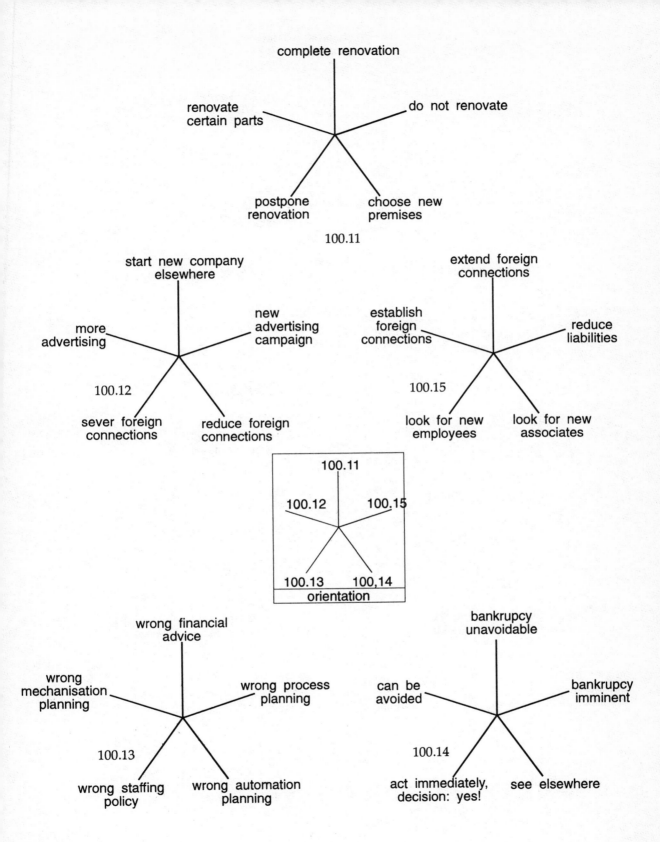

complete renovation

renovate certain parts — do not renovate

postpone renovation — choose new premises

100.11

start new company elsewhere

more advertising — new advertising campaign

100.12

sever foreign connections — reduce foreign connections

extend foreign connections

establish foreign connections — reduce liabilities

100.15

look for new employees — look for new associates

100.11

100.12 — 100.15

100.13 — 100,14

orientation

wrong financial advice

wrong mechanisation planning — wrong process planning

100.13

wrong staffing policy — wrong automation planning

bankrupcy unavoidable

can be avoided — bankrupcy imminent

100.14

act immediately, decision: yes! — see elsewhere

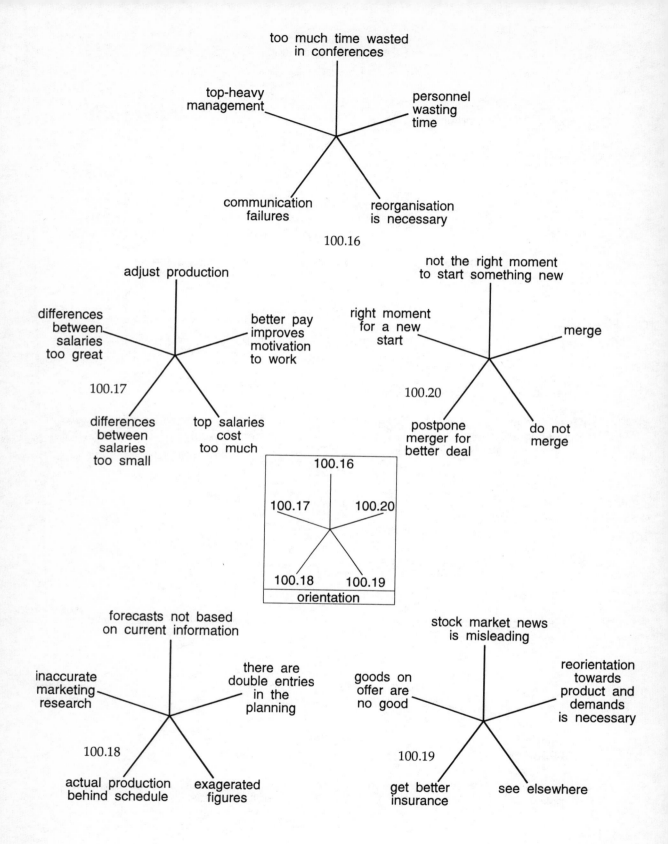

too much time wasted
in conferences

top-heavy
management

personnel
wasting
time

communication
failures

reorganisation
is necessary

100.16

adjust production

differences
between
salaries
too great

better pay
improves
motivation
to work

100.17

differences
between
salaries
too small

top salaries
cost
too much

not the right moment
to start something new

right moment
for a new
start

merge

100.20

postpone
merger for
better deal

do not
merge

100.16

100.17   100.20

100.18   100.19
orientation

forecasts not based
on current information

inaccurate
marketing
research

there are
double entries
in the
planning

100.18

actual production
behind schedule

exagerated
figures

stock market news
is misleading

goods on
offer are
no good

reorientation
towards
product and
demands
is necessary

100.19

get better
insurance

see elsewhere

168

22. Lost objects and missing persons

This chapter is for people who have lost touch with certain friends, acquaintances or relations. Only use the pendulum to make sure about the whereabouts of people you know. Do not try to trace missing politicians or other public figures just to satisfy your thirst for sensation. Such improper use of the pendulum is not recommended. People who are elsewhere can be located by means of the diagrams in chapter 16. You can also use the pendulum for missing objects. Often someone is able to remember when or where the object was seen last.

To simplify matters you can use the following diagrams.

Once the pendulum has indicated an area, a detailed drawing might be useful. It will enable you to pin-point the spot more accurately.

Lost pets are another problem. They can be anywhere and are often walking about, which makes searching even more difficult. Again, you may locate the exact position with the help of a detailed map. But a phone call or visit to the home for lost animals or police station may be just as useful.

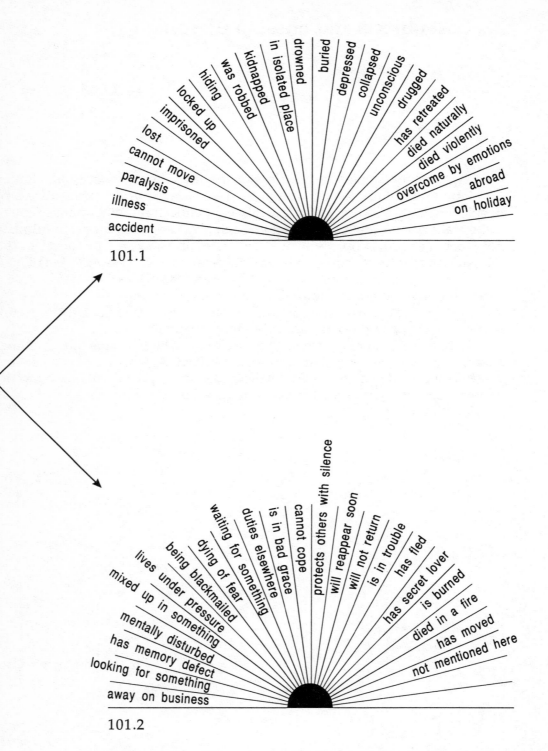

101.1

101.2

Figure 101. What has happened to a missing person

170

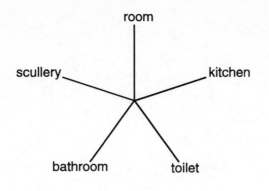

room

scullery ╱ kitchen

bathroom ╲ toilet

102.1

cupboard

tool shed ╱ outdoors

102.2

garage ╲ storage room

stable

veranda ╱ garden house

102.5

hall ╲ loft

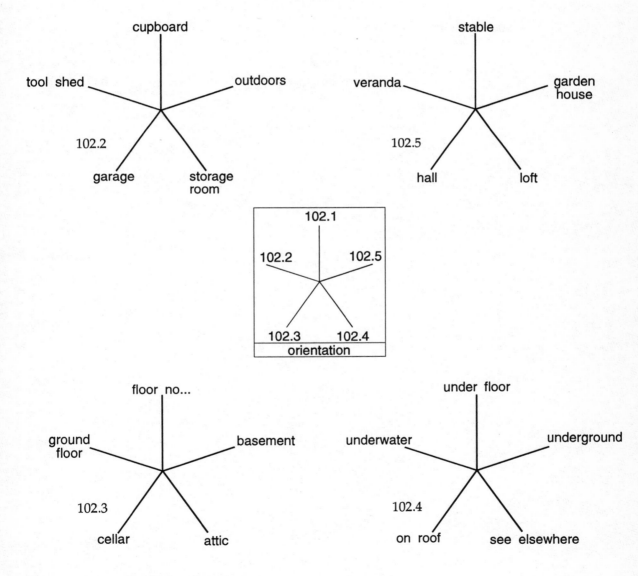

102.1

102.2 ╱ 102.5

102.3 ╲ 102.4

orientation

floor no...

ground floor ╱ basement

102.3

cellar ╲ attic

under floor

underwater ╱ underground

102.4

on roof ╲ see elsewhere

Figure 102. Location

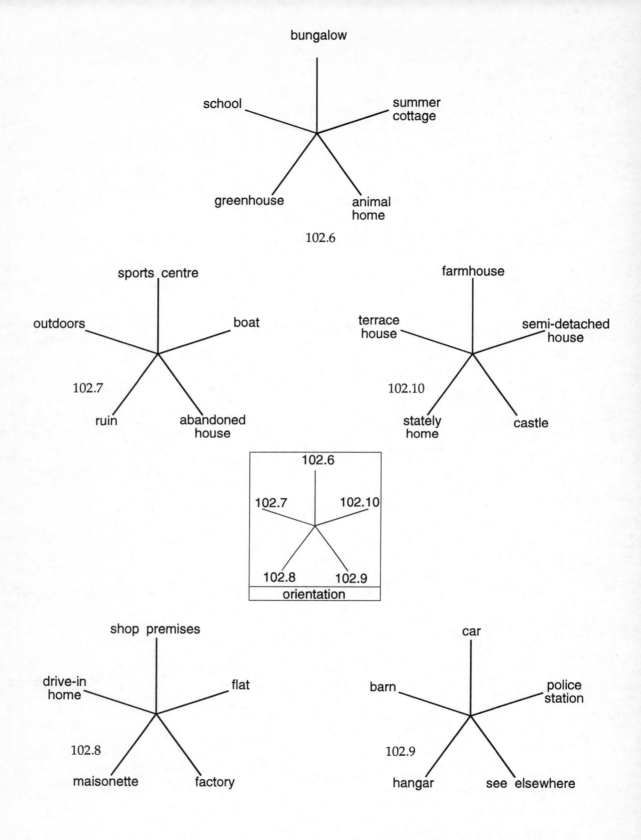

bungalow

school

summer cottage

greenhouse

animal home

102.6

sports centre

outdoors

boat

102.7

ruin

abandoned house

farmhouse

terrace house

semi-detached house

102.10

stately home

castle

102.6

102.7   102.10

102.8   102.9

orientation

shop premises

drive-in home

flat

102.8

maisonette   factory

car

barn

police station

102.9

hangar   see elsewhere

172

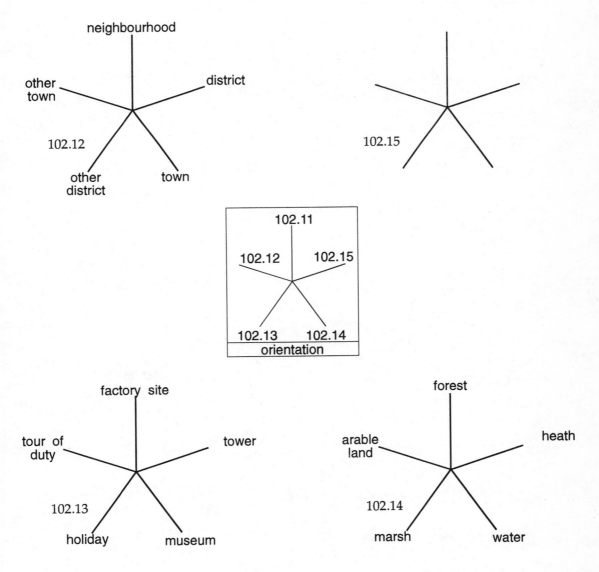

school

garden church

workshop garage

102.11

neighbourhood

other town district

102.12

other district town

102.15

102.11

102.12 102.15

102.13 102.14

orientation

factory site

tour of duty tower

102.13

holiday museum

forest

arable land heath

102.14

marsh water

cupboard

car   desk

lost and found   vault
department

102.16

stolen

lent out   robbed

102.17

drain   sewer

tram

bus   train

102.20

102.6

102.7   102.10

102.8   102.9
orientation

in bed

changing   under
room   bed

102.18

canteen   lost while
             exercising

lost while
travelling

lost while   at work
shopping

102.19

lost at   see next page
home

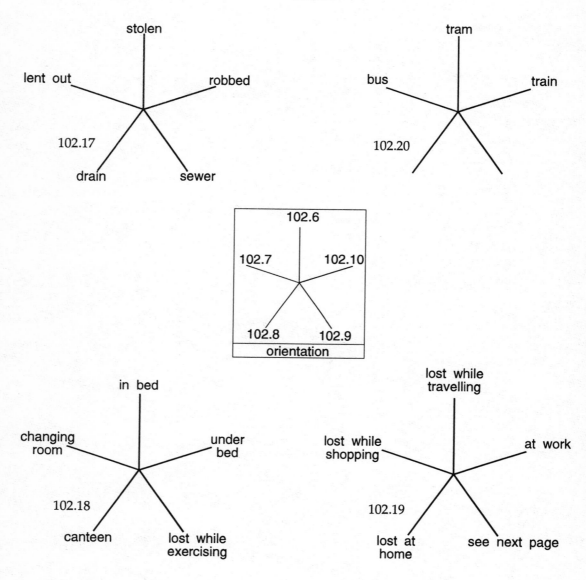

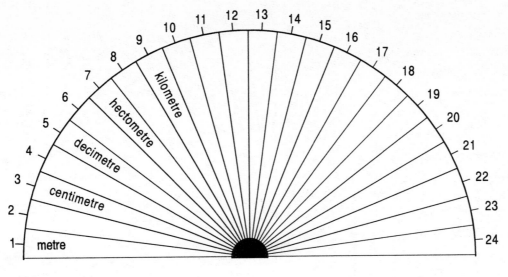

103.1

higher   exact   lower

103.3

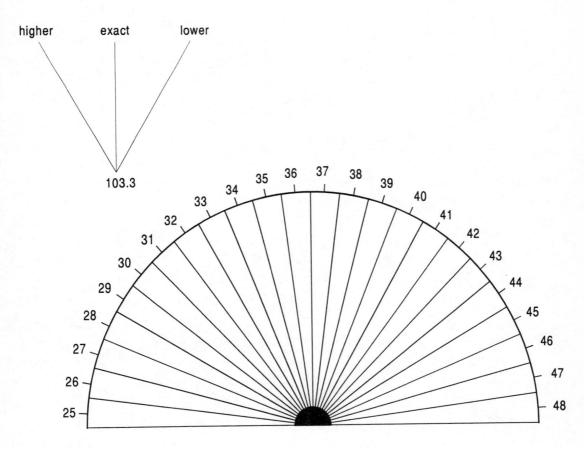

103.2

Figure 103. Distance to lost object; height or depth

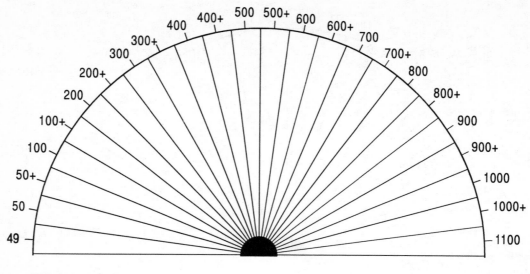

103.4

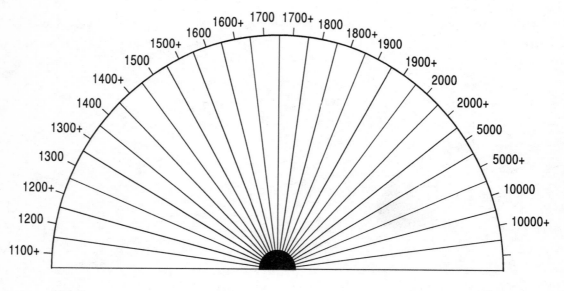

103.5